Materials Science
WATER

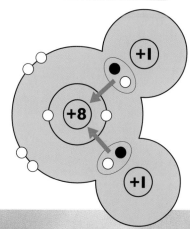

making use of the secrets of matter

Atlantic Europe Publishing

First published in 2003 by
Atlantic Europe Publishing Company Ltd.

Author
Brian Knapp, BSc, PhD

Art Director
Duncan McCrae, BSc

Senior Designer
Adele Humphries, BA, PGCE

Editors
Mary Sanders, BSc, and Gillian Gatehouse

Illustrations
David Woodroffe

Design and production
EARTHSCAPE EDITIONS

Scanning and retouching
Global Graphics sro, Czech Republic

Print
LEGO SpA, Italy

Materials Science – Volume 8: Water
A CIP record for this book is available from the British Library

ISBN 1 86214 322 6

Acknowledgments
The publishers would like to thank the following for their kind help and advice: *Thames Water plc (Farmoor Advanced Water Treatment Works).*

Picture credits
All photographs are from the Earthscape Editions photolibrary except the following: (c=center t=top b=bottom l=left r=right)
London Fire Brigade 36b; UKAEA Technology 15.

This product is manufactured from sustainable managed forests. For every tree cut down, at least one more is planted.

Contents

(*Left*) Interlocking ice crystals on the surface of freezing water.

1: Introduction

A MATERIAL is something that people work with to make something that others need. Anything that is SOLID, LIQUID, or GAS can be a material.

Most materials that we want to use have to be taken from the earth. Rocks have to be heated to release their metals; plastics come from oil obtained by deep drilling. But water is different—it is all around us. All we have to do is collect it from rivers and the sea.

The universal presence of water has been noted since ancient times. Some ancient Greeks (for example, Thales of Miletus) believed that water was the only basic building block of matter and that everything else was formed from it.

Another Greek, Aristotle, believed water to be one of four fundamental "elements," the others being earth, air, and fire. It took until the 18th century to show that this was not so and that water is made from oxygen and hydrogen.

2. Clouds

There is water vapor in the air around you now, but it is so thinly spread that you cannot see it. However, high in the cold clouds moisture forms into tiny droplets, which then grow large enough to fall from the air as rain (or snow if the air is very cold).

1. The oceans

More than nine-tenths of the world's water is in the oceans. However, we can't drink this water because it contains lots of salt. Liquid water changes to invisible water vapor due to the warmth of the air and the heat from the Sun. This is known as EVAPORATION.

(Right) The natural water cycle. Notice that water changes from a liquid to a gas (vapor) and back to a liquid. It is never used up.

The water vapor that rises from the oceans contains no salt. It will form the fresh water that we can drink.

The water cycle

Why is water so readily available? It is because it constantly circulates between the oceans, the air, and the land (including the rocks) in a pattern known as the WATER CYCLE. We are all so familiar with it that we tend to take it for granted. But the water cycle only works because water can occur in the three forms, or states: solid, liquid, and gas, at normal temperatures. We will find out more about the importance of states of matter on pages 21 to 23.

3. Rain
Rainwater is pure water—it is not salty and normally quite safe to drink. But it is not easy to catch. When rain falls, most of it immediately sinks into the soil. However, some other living things are better able to use it than us. Most important are plants, which suck up water from the soil through their roots.

4. Rocks
Any rain not used by plants seeps into the ground, first through the soil and then into the rocks below. Water moves very slowly through the rocks. It may take weeks or months to travel to a river.

5. Rivers and lakes
Once the rocks are full, water begins to seep out at the surface, perhaps as springs, mostly unseen through the beds of rivers and lakes. This (not rainfall) is the main source of water for most rivers. Finally, rivers carry the water back to sea and so complete the water cycle.

6. Floods
Only during the heaviest rain or after the wettest of seasons do we ever see water flowing on the surface. When it does this, we call the surface water a flood.

Water is not evenly distributed as a global resource. Over 97% of it is in the oceans and just over 2% more in the ice caps. That leaves just about 1% for lakes, rivers, rocks, soils, and the air. But when used carefully, that is more than enough for the needs of all living things.

A transporting medium

One of the main things that water does is to carry other materials with it. Some of this material is in SOLUTION (see below), but some also is in SUSPENSION. This is extremely common. For example, blood is a suspension of blood cells in water. On a larger scale sewage is transferred from homes to sewage plants in suspension.

A chemical

Water is extremely abundant and also very chemically REACTIVE. As a result, it is the most important and most common of all chemical COMPOUNDS on the Earth's surface.

Water is called the "universal SOLVENT." A vast array of substances is DISSOLVED in it naturally, which is why we have salty ocean water and "hard" fresh water that is difficult to lather. It is also a splendid means of getting chemicals to large numbers of living things cheaply, which is why fertilizers are put in irrigation systems on farmland and why fluoride is added to many drinking water supplies to protect whole populations from tooth decay.

Because it is such a good solvent, water is rarely pure. Pure water can be prepared in a laboratory by boiling it and collecting the water vapor that is given off. The closest parallel to this in nature is evaporation. But water vapor in the air is not often pure because it can dissolve or react with other gases in the atmosphere. It can form weak acids that can be damaging to trees and animals.

If you follow the water cycle, you can also see that it is a chemical scavenging cycle too. Once water

(Above) Our drinking water is not pure. It contains many dissolved substances, some of which are beneficial and others harmful. It is the job of water companies to help produce a combination that is safe to drink.

forms into droplets (which make clouds), it has already attracted gases from the air and has become weakly acidic. Water droplets form around specks of dust in the air. Many such specks are made of SALT, which are then dissolved in the water.

The water that contains the greatest impurities is the ocean and some water in rocks (known as GROUNDWATER). That is because water reacts with many rocks, and some of the products of the REACTIONS remain dissolved in the water.

Water moving through the ground is always in contact with rocks, and so it can dissolve considerable quantities of MINERALS. Indeed, this kind of water is often valued for its impurities. It is called "mineral water" and is believed by some to have health-giving properties. The sea contains salts that have been dissolved in rocks and carried by rivers. When the water evaporates to form more water vapor and so continue the water cycle, it leaves behind its impurities, leading to "salty" water, although salt (sodium chloride) is not the only substance dissolved in seawater.

Water is also an excellent CATALYST for many chemical reactions, and it can act either as an OXIDIZING AGENT or as a REDUCING AGENT. For example, water can oxidize carbon to carbon monoxide, liberating hydrogen gas. It can also reduce chlorine gas to hydrogen chloride, releasing oxygen gas.

(*Below*) The substances dissolved in water are clear to see in dried-out lake beds such as this one in a desert region of California.

(*Above*) The ice shelves around Antarctica and Greenland are simply thick slabs of ice floating on water. They act as a protective blanket for the water underneath.

(*Above*) If water did not have the strange property of being lighter as a solid than a liquid, then ice fishing would not be possible.

Physical properties of water

Many of the properties of water are very unusual and some quite unique. For example, although most solids are more dense than their liquids, ice is less dense than liquid water. That is why ice cubes float in a glass of water. Liquid water is most dense at 4°C, not at its FREEZING POINT, as is the case for other liquids.

Many such properties are vital to the way our world works, the way that water is used by living things, and the way we use water in homes and factories. For example, if ice did not float but was heavier than liquid water, it would sink to the bottom of a lake as soon as it formed on the surface. In cold conditions more and more water would turn to ice and sink, and eventually the lake would fill with ice. This ice would also be very slow to melt because it would be protected from warmth by a layer of surface water. Under these conditions the Great Lakes, for example, would be the Great Ice Blocks.

Water-living creatures also depend on ice floating. The ice forms a protective blanket on the surface of the lake. At the same time, as water cools to 4°C, it sinks to the bottom of the lake. In this way, except in the most unusual of circumstances, although the surface of a river, lake, or ocean freezes, the bottom stays unfrozen and provides a sheltered home to living things.

The nature of water molecules

Many of the properties of water outlined above are the result of the special nature of water MOLECULES. Water, or hydrogen dioxide, consists of two hydrogen ATOMS bonded to one oxygen atom. That is why it is represented by the formula H_2O. As it happens, this linking of hydrogen and oxygen leaves both hydrogen atoms on one side of the molecule and the oxygen on the other side. The atoms are BONDED at an angle of about 104 degrees. The hydrogen atoms are attached to the oxygen by COVALENT BONDS.

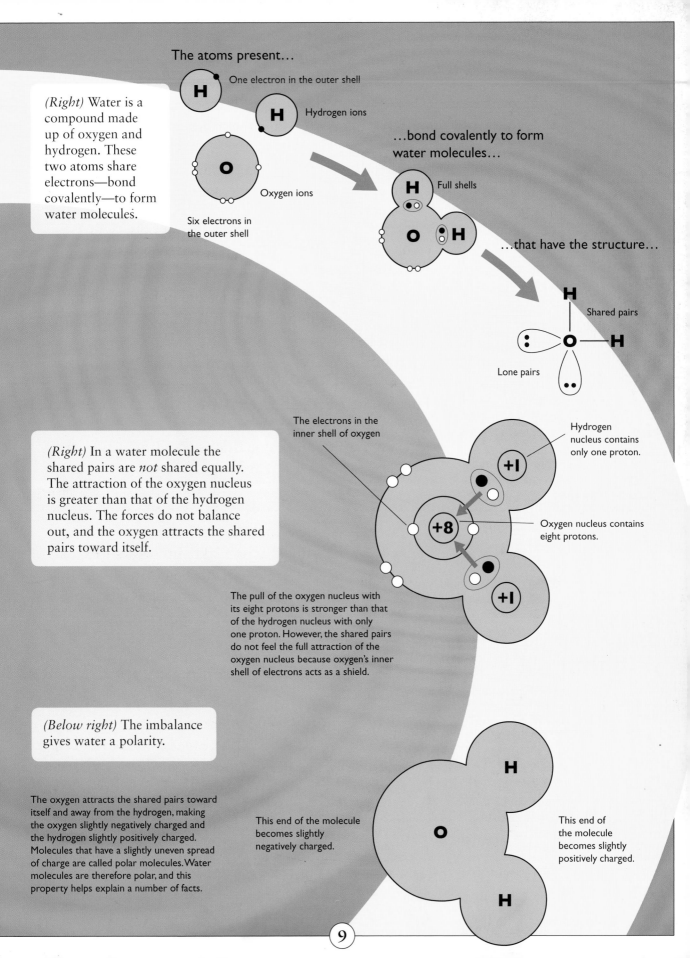

The atoms present...

One electron in the outer shell!

Hydrogen ions

Six electrons in the outer shell

Oxygen ions

...bond covalently to form water molecules...

Full shells

...that have the structure...

Shared pairs

Lone pairs

(Right) Water is a compound made up of oxygen and hydrogen. These two atoms share electrons—bond covalently—to form water molecules.

(Right) In a water molecule the shared pairs are *not* shared equally. The attraction of the oxygen nucleus is greater than that of the hydrogen nucleus. The forces do not balance out, and the oxygen attracts the shared pairs toward itself.

The electrons in the inner shell of oxygen

Hydrogen nucleus contains only one proton.

Oxygen nucleus contains eight protons.

The pull of the oxygen nucleus with its eight protons is stronger than that of the hydrogen nucleus with only one proton. However, the shared pairs do not feel the full attraction of the oxygen nucleus because oxygen's inner shell of electrons acts as a shield.

(Below right) The imbalance gives water a polarity.

The oxygen attracts the shared pairs toward itself and away from the hydrogen, making the oxygen slightly negatively charged and the hydrogen slightly positively charged. Molecules that have a slightly uneven spread of charge are called polar molecules. Water molecules are therefore polar, and this property helps explain a number of facts.

This end of the molecule becomes slightly negatively charged.

This end of the molecule becomes slightly positively charged.

Because it is made of covalent molecules, water is a very poor CONDUCTOR of electricity. But at the same time, that is why water is an excellent solvent, and many substances that are IONIC will dissolve in it. This works because the water dramatically reduces the attractive forces between the CATIONS and ANIONS of a substance and allows them to break apart (see page 44).

Many of the properties of water result from its very unusual structure. For example, a water molecule is highly POLAR. That means it behaves as though it has a positive charge at one end and a negative charge at the other. The poles are found at the oxygen atom (which has a net negative charge) and midway between

(Below) A rod charged with static electricity will attract a stream of water. The polar water molecules line up and are attracted to the opposite charge in the rod.

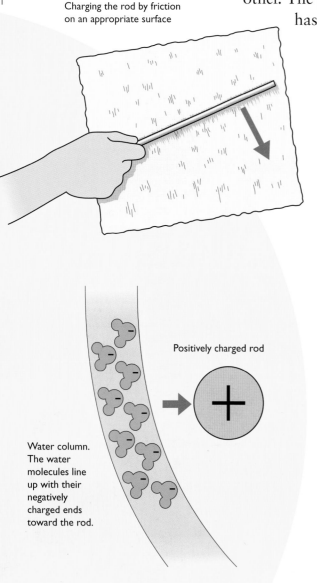

Charging the rod by friction on an appropriate surface

Positively charged rod

Water column. The water molecules line up with their negatively charged ends toward the rod.

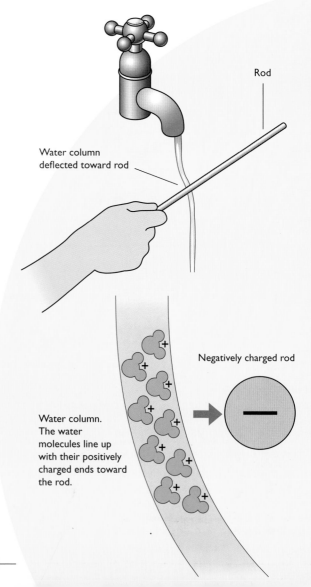

Rod

Water column deflected toward rod

Negatively charged rod

Water column. The water molecules line up with their positively charged ends toward the rod.

the hydrogen atoms (which have a net positive charge). This polar property is produced from the bent shape of the molecule. The polarity (charged nature) can be demonstrated by charging up a plastic comb with static electricity (rubbing it against clothing) and then placing it close to a thin stream of water flowing from a faucet. The stream of water will bend toward the comb, attracted by static electricity.

Hydrogen bonding

Water molecules bond to one another in special ways. One of them is called HYDROGEN BONDING. Hydrogen bonding of water molecules gives water many properties quite unlike any other substance. It accounts for the SURFACE TENSION (see page 25) effect in water as well as causing MELTING POINTS (see page 17) and boiling points (see pages 14 and 20) higher than would be expected if water had properties common to other substances.

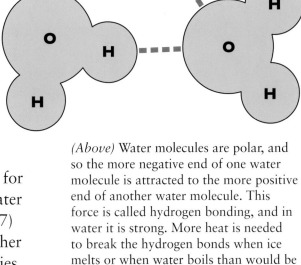

(Above) Water molecules are polar, and so the more negative end of one water molecule is attracted to the more positive end of another water molecule. This force is called hydrogen bonding, and in water it is strong. More heat is needed to break the hydrogen bonds when ice melts or when water boils than would be needed if the molecules were not polar. That is what makes water's fixed points of boiling and melting so high.

Hydrogen bonds occur throughout the liquid and solid states of water.

(Below) The surface tension effect of water is helped by hydrogen bonds.

For example, when water freezes into ice, hydrogen bonding holds the water molecules apart in a LATTICE that makes ice only nine-tenths as dense as liquid water. Each water molecule of an ice crystal is further bonded to its neighbors by weak hydrogen bonds. That is why water expands to ice (turns from liquid to solid) quite unlike other liquids, which keep the same volume or even shrink slightly as they solidify. There are far more hydrogen bonds in ice than in liquid water; and as melting occurs, some of the hydrogen bonds break, allowing water molecules to pack together more closely and making liquid water much denser than the solid.

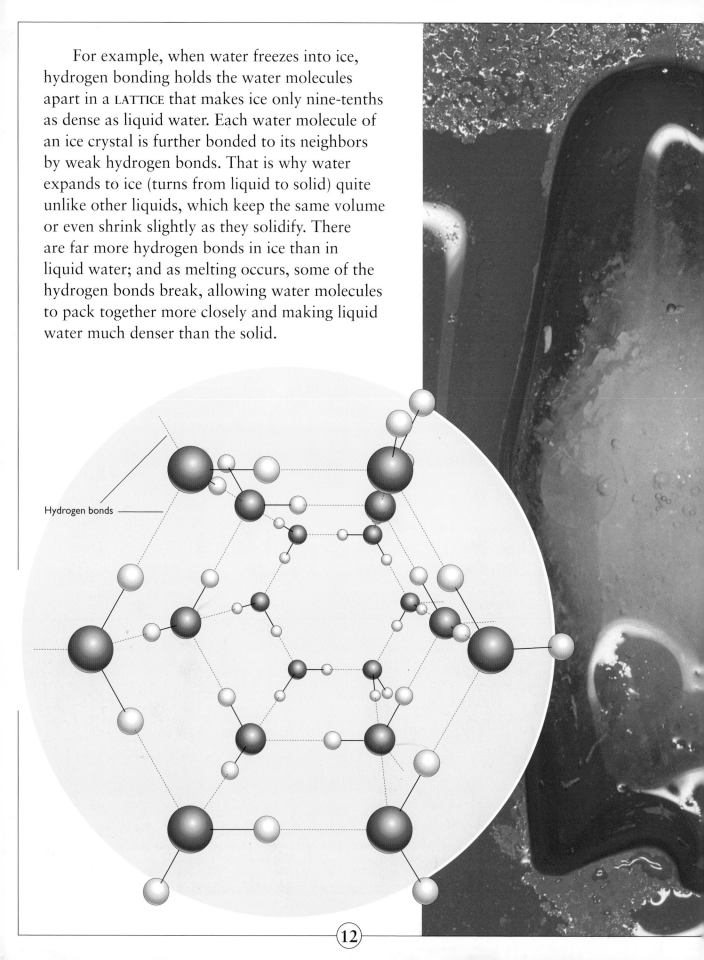

Hydrogen bonds

(Left and below) The open structure of ice makes it less dense than liquid water. Each water molecule has only four nearest neighbors with which it interacts by means of hydrogen bonds. The shape of the rings is mirrored in the sixfold symmetry of snowflakes.

The increased volume of ice compared with water is important in rock weathering, where the expanding ice on a wet surface levers loose blocks from the surface.

The forces involved as ice expands can easily crack a bottle. If plant cells with relatively fragile cell walls are frozen, they can rupture. This can damage crops both before they are harvested and during transit or storage.

The effect of linking molecules is very important. Normally, a liquid made of small molecules flows freely and boils easily. But water is not like that. The bonding makes water "thicker" or "stickier" (more VISCOUS) than it would otherwise be. It also means that the surface holds together as though it had an invisible skin (called surface tension). It also boils at a much higher temperature (100°C) than would otherwise be expected (without the bonding it would boil at −100°C!). So water would not exist as solid or liquid on Earth at all and would be more like carbon dioxide or oxygen gases.

Radioactive water

Many elements can exist in different forms called ISOTOPES. Sometimes the isotopes are RADIOACTIVE. Isotopes of hydrogen include $_2$H, or D, deuterium, and $_3$H, or T, tritium. These isotopes combine with oxygen and form water molecules. Heavy water is D_2O, deuterium oxide. Heavy water and tritium can be combined to release enormous amounts of energy and can be used in the making of nuclear bombs.

Deuterium can also be used to slow down the neutrons produced in a nuclear reaction. A material that can do this is called a "moderator." Heavy water can be used as a moderator. Heavy water is a poor absorber of neutrons, and that allows heavy water reactors to use natural (nonenriched) uranium as a fuel.

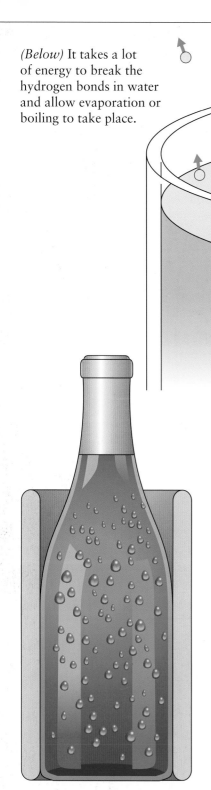

(Above) The thermal capacity of water partly relates to its hydrogen bonds. The energy used in melting the ice in this ice cooling jacket partly comes from the bottle it encloses. This cools the bottle.

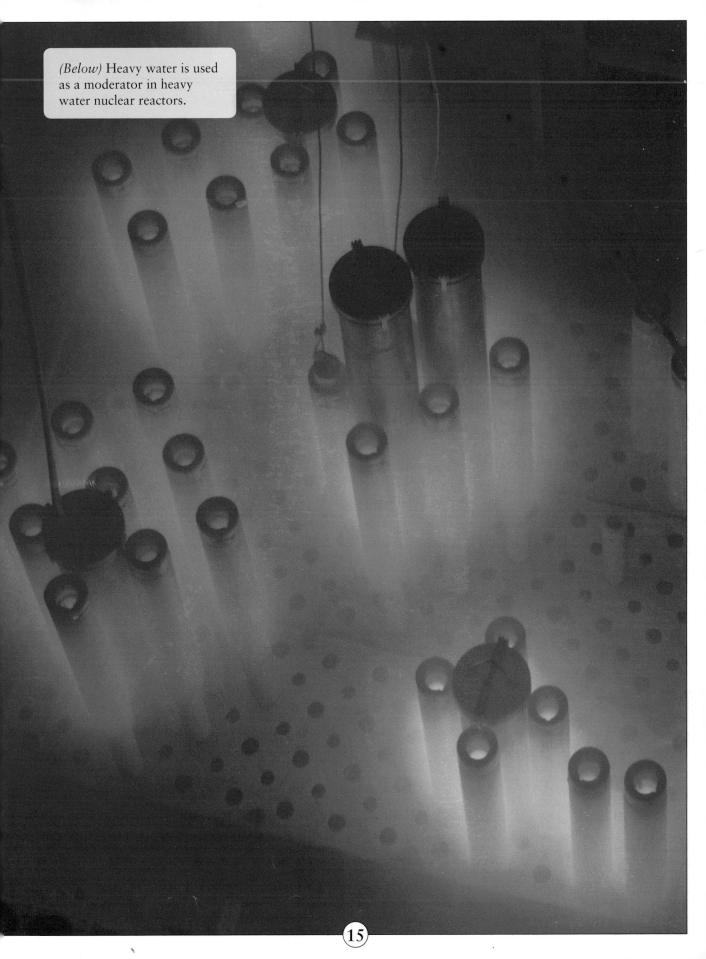

(Below) Heavy water is used as a moderator in heavy water nuclear reactors.

2: States of water

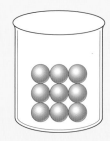

Many compounds can be a solid, a liquid, or a gas. Each of these forms is called a STATE OF MATTER.

We call the liquid form of water simply "water" because that is how it normally appears to us. The other states of water are as a gas—called WATER VAPOR (see pages 19 and 20)—and as a solid—called ice. Water can also take part in many chemical reactions, that is, it is a reactive substance.

STEAM, which is water vapor at the boiling point of water, is a particularly reactive substance; liquid water is still quite reactive, but ice is inert and unreactive.

The change from steam to water leads to CONDENSATION and a reduction in volume; the change from water to ice, SOLIDIFICATION, leads to an expansion in volume. We will see more of this below.

Water (a molecule of hydrogen and oxygen atoms) may be reactive, but it is also a stable substance and will not DISSOCIATE into hydrogen and oxygen gases unless a large amount of energy is applied, for example, in the form of an electric current (see pages 55 and 56). Thus, heating water to its boiling point will not make it dissociate; it merely makes it VAPORIZE.

As we have seen, water can form a gas, a liquid, and a solid. But not all substances are like this. That is partly because water consists of two atoms (hydrogen and oxygen), while some liquids

(Right) The attractive forces between particles are different for each state.

See **Vol. 9: Air** *for more on water vapor.*

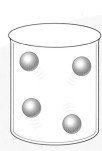

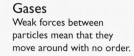

Gases
Weak forces between particles mean that they move around with no order.

Evaporation (vaporization)

Condensation

Sublimation

Sublimation

Solidification (freezing)

Melting

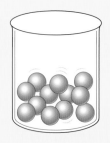

Liquids
The forces between particles in a liquid are intermediate between gases and solids and lead to a loose, irregular arrangement.

Solids
Strong forces between particles in a solid give a closely packed and regular structure. There is little space between particles.

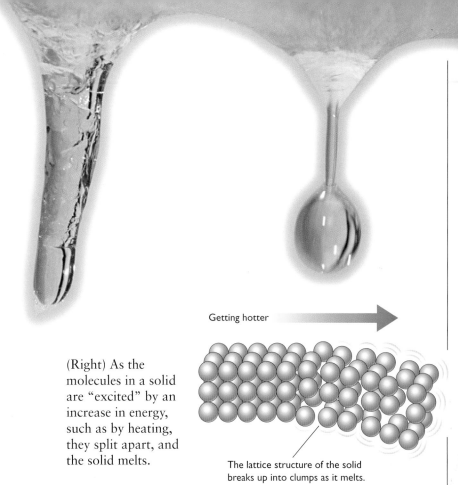

(Above and above right)
Freezing and the development
of ice crystals and melting.

(Below) Sometimes water can
change directly from a gas
to a solid without forming a
liquid in between. This is called
sublimation, and it is shown when
ice forms on cold surfaces. We
then call it hoarfrost. Under strong
heating ice can also evaporate,
forming water vapor without
becoming a liquid in between, and
this too is called sublimation.

(Right) As the
molecules in a solid
are "excited" by an
increase in energy,
such as by heating,
they split apart, and
the solid melts.

Getting hotter

The lattice structure of the solid
breaks up into clumps as it melts.

(such as latex) consist of hundreds or thousands
of atoms. Although they can be both liquids and
solids, they can never be gases. When they are
heated, the atoms all split apart, and the liquid
DECOMPOSES instead of evaporating.

Changing between states

For a liquid to change to a gas or solid, there must
be a change in its energy. For example, liquid water
possesses enough energy to stop the attractive forces
between molecules from fixing it into the rigid solid
structure (ice), but it does not have enough energy
to allow each molecule to move away from its
neighbors and form a gas (water vapor). So, when
a liquid changes into a gas, it must gain some extra
energy; and when it changes from a liquid to a solid,
it must lose some energy. The energy gain or loss is
called LATENT HEAT.

Evaporation and condensation

When water is exposed to the air, such as over the surface of a river, lake, or ocean, some molecules are always escaping and turning into gas. This is evaporation.

Evaporation occurs because molecules do not all soak up the same amount of heat. A few molecules are always getting enough heat to break away from their neighbors and turn into a gas, just as a few molecules of gas are always losing energy and slowing down enough so they get captured by the liquid.

Clearly, the more heat there is available, say from sunshine, the greater the chances are that more water molecules will get enough heat to break free than those losing energy and turning into water.

The molecules that escape become part of the air and also help increase the weight (pressure) of the air on the surface.

When air is as full of water vapor as possible, then, on balance, no more water can leave the liquid. The air is then said to be SATURATED.

(Above) Condensation is the reverse of evaporation and so occurs where water vapor loses energy and turns into liquid. This often happens on a cold surface such as a window or a glass of cold liquid.

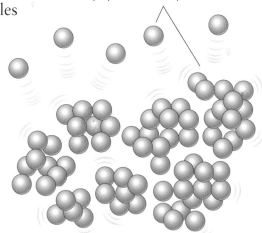

Clumps of particles move rapidly, and some escape.

(Above) Evaporation is the loss of some molecules of water from the surface of the liquid.

When a liquid evaporates, it does so without any visible sign. That is why, if you leave a saucer with water in it, all the water will eventually disappear almost mysteriously.

As heat is taken away, as on a cool night, the opposite is true, with more gas molecules losing heat and so slowing down than liquid molecules gaining heat and breaking away. The result is the coating of water we find on cold surfaces early in the morning. It is condensation.

In a pure liquid, such as pure water, the molecules of the gas are exactly the same as the molecules in the liquid. But in a solution (which is a mixture), this is not true, and the vapor contains a greater concentration of the molecules that find it easiest to evaporate. When seawater is heated, the water molecules evaporate, but the salt remains behind because the water molecules evaporate far more easily than the salt.

(Right) The gaseous form of water is called water vapor. Evaporation provides enough energy for water to leave the liquid state and become gaseous. Steam is gaseous water at boiling point or above.

Steam is an invisible gas. The white "steam" we can see coming from a kettle as it boils is, in fact, tiny droplets of water that have condensed as the steam moves into cold air. The water vapor has condensed into its liquid state, which is why it will wet nearby surfaces. But if you look carefully at a kettle boiling, you will see that just beyond the spout of the kettle there is a zone with no droplets. This is pure steam (that is, water as an invisible gas).

Liquid again—water vapor condensing as water droplets that we can see.

Water vapor—gas; we can't see the vapor.

Water boiling inside the kettle

Boiling

When water boils, the water gets enough heat energy for molecules to push apart *inside* the water as well as on the surface. That is why bubbles form inside the liquid. The bubbles do not contain air—they contain only water vapor. We recognize bubbles forming inside the water as a sign of boiling. Boiling is thus a very fast form of evaporation.

(Left, right, and below) Boiling is the change from liquid to gas within the water as well as at the surface. That is what causes it to bubble. Each bubble contains water vapor (not air).

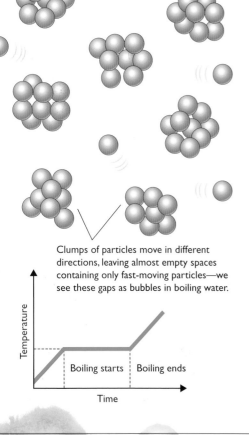

Many of the particles can now escape from the liquid surface.

Clumps of particles move in different directions, leaving almost empty spaces containing only fast-moving particles—we see these gaps as bubbles in boiling water.

Temperature

Boiling starts | Boiling ends

Time

Water vapor occupies much more space than liquid water. So, as liquid turns to vapor, bubbles swell. They are also less dense than the liquid, and so they rise through the liquid.

The higher the bubbles rise, the less weight of water lies above them, so the less pressure there is on the vapor. As a result, the vapor expands, and the bubbles grow as they approach the surface. When they get to the surface, the bubbles then burst, sending water vapor (not air) into the air above.

It is possible to keep water very tightly confined. If this is done, the water cannot boil, and so it becomes superheated, and its temperature rises above normal boiling point. When the pressure is released, steam is produced very violently. It is also possible to produce steam and keep it confined so that it cannot expand and then allow it to expand. Again, the change in volume occurs very quickly.

Making use of changes of state

The property that water will change easily from solid to liquid to gas is extremely important. In fact, a large part of the INDUSTRIAL REVOLUTION, which changed us from an agricultural society to an industrial one, depended on the change of state of water, especially the change from liquid to gas and back again.

Steam engines

Behind all of the early industrial machines was the knowledge that water could be changed from liquid to gas (in the form of steam) and then back to water again.

In the late 17th century Denis Papin invented a pressure cooker, that is, he enclosed water in a container with a tightly fitting lid and then heated it. In this way he created steam under pressure, and the result was that it blew the lid off the container. From this came the idea of using steam to power a piston and cylinder engine.

One of the earliest machines to use this principle was the pumping engine. The pumping engine was developed to draw water out of deep coal mines.

In a simple steam pump water is heated in a boiler until it is turned to steam by heating the water well above its boiling point. In this way the steam is under pressure. The steam is then sent into a cylinder containing a piston.

Expanding steam moves very fast and with great force (as heat energy is converted into mechanical energy of movement). For example, expansion from a pressure of about 12 times atmospheric pressure down to about half atmospheric pressure occurs with a speed of about 1,100 meters per second.

(Above and below) Pumping engines for mines were the first steam engines.

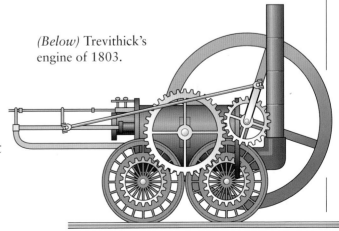

(Below) Trevithick's engine of 1803.

(Right) In a steam engine water is heated by a furnace to form steam. The water is held in a boiler so that as it is converted to steam, it tries to expand. But because it is confined by the boiler, there is a buildup of pressure.

The pressurized steam is fed to the cylinders, where it presses on a piston, causing it to move along a cylinder. By feeding steam alternately into the cylinder at both ends, the piston can be moved back and forth, and a simple CRANKSHAFT can be fitted to drive the wheels.

(Since the fuel used in a steam engine is burned outside the cylinder, it is called an external combustion engine.)

High-pressure steam fed into the cylinder

Furnace used to heat the water

Cylinder

Piston

In the boiler water is heated and turned into steam.

The pressure of the expanding steam causes the piston to be pushed down the cylinder. The steam expands as it enters the cylinder, cools (or is cooled very quickly by using cold water), and turns into liquid. The change in volume as the gas changes state to liquid produces a vacuum in the cylinder. Atmospheric pressure working on the other end of the piston, coupled with a vacuum inside the cylinder, then draws the piston back up the cylinder.

In this way changes of state can be used to produce the back-and-forth motion of a piston in a cylinder. By attaching a suitable rotating crank on a

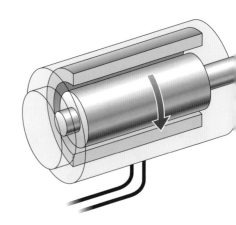

flywheel to the bottom of the piston, it can be turned into a rotational movement, which can turn the wheels of a lathe, a loom, a tractor, a steam engine, or many other machines. It is both simple and reliable.

A more sophisticated form of steam engine (double-acting engine) lets in steam to each side of the piston in turn so that simple vacuum draw back is aided by more steam pressure. A series of valves controls the entry of steam at appropriate moments.

Steam turbines

Steam turbines are more modern versions of the use of steam to drive a machine. In the steam turbine high-pressure steam comes at high speed through nozzles and then presses against a series of blades attached to a shaft (something like a propeller in reverse). In fact, the best power is achieved by sending the steam across a series of both fixed and moving blades.

(Below) Even today, steam is at the heart of energy generation. Gas, coal, and nuclear energy are all used to heat water and generate the steam that drives generators.

Steam turbines are huge and use high-pressure steam to drive complex turbine blades that turn a shaft that drives generators to make electricity.

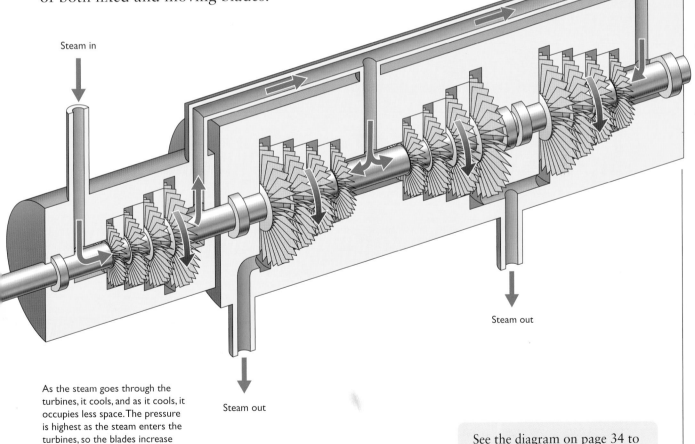

Steam in

Steam out

Steam out

As the steam goes through the turbines, it cools, and as it cools, it occupies less space. The pressure is highest as the steam enters the turbines, so the blades increase in size along the turbine as the pressure gets progressively lower.

See the diagram on page 34 to see how a steam turbine such as this is used in a power plant.

3: Physical properties of liquid water

(Left) Water will flow and settle to fill the bottom of a container no matter what its shape.

Liquids move easily. So if you get a glass of water and tilt it, the liquid will first change shape and then pour from the glass. Notice that the water has not changed its volume, just its shape. Notice, too, that the upper surface of the water is flat—all of the change in shape occurs where the water touches the glass.

Once you pour the water from a glass into, say, a dish, the water then takes up the shape of the dish. While it is pouring, the liquid tries to hold together. However, if air resistance causes the water to break up into droplets, as soon as each droplet gets to the dish, it will immediately join with the others to form a whole again.

Water behaves as though it were a collection of tiny slippery balls that could move freely, but that, at the same time, are held together by a force. In this case the tiny balls are called molecules. This is very

(Below) Water can occur as three phases—gas, liquid, and solid.

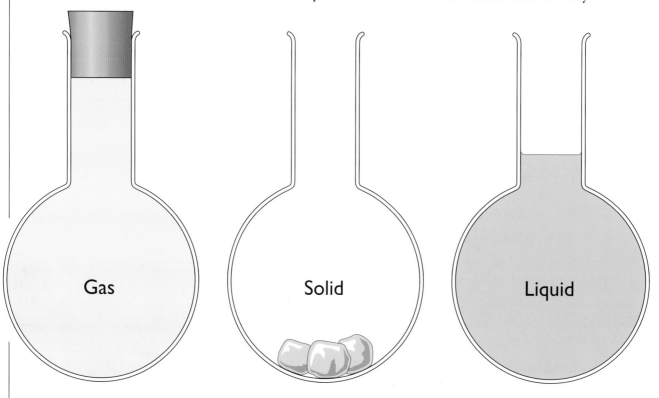

Gas

Solid

Liquid

different from a gas or a solid. We could not have kept a gas in an open glass because a gas expands to fill the shape of the container it is in, and if there is no lid, it will simply expand out into the air. If we tilted a glass with solid water in it, the ice blocks might tumble around, but they would not change their shape. And when they were poured out, they would not re-form into a single whole.

Surface tension

The molecules in water can move around even though they are attracted together, and that is why water has no fixed shape. Yet because the molecules are all bound to one another, they form a heavy mass that collectively sinks down to the bottom of any container it is placed in.

Water is held together by forces that are balanced inside the liquid. But at the surface water molecules are more attracted to their neighbors than they are to the molecules in the air. Thus there is a dominant sideways force and an inward force, but there is no outward force. This imbalance creates a net inward force, and it is this force that makes water form a shape that has the smallest possible surface area and thus behave as though it were surrounded by an invisible skin.

We know this surface force as surface tension. We also see it, for example, as a faucet drips or when water globules gather on the side of a cold glass. It is also the force that holds raindrops together. You can see this effect used by tiny water insects such as pond skaters. Their light weight means they do not break the force between the molecules, and so they merely dent the surface where they stand.

The surface tension effect works on all liquids, but it is particularly strong in the case of water. The strength of the surface tension force explains why water tries to gather into little balls instead of spreading out over the surface it is resting on.

In places where gravity does not work (for example, in spacecraft) water droplets in air are truly spherical. This also shows that surface tension is a property of the water and is nothing special to the Earth. But in the Earth's gravity the weight of the liquid causes the shape to

(Below) Surface tension attracts the surface water molecules together, holding them together as though they had an invisible skin.

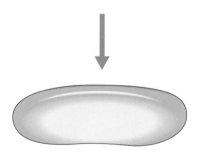

change. That is why most droplets on a table become flattened, and why when they fall through the air, they are not teardrop shaped (as popularly drawn) but become almost doughnut shaped.

Capillarity

When water enters very small spaces, such as the gaps in a sponge or the bore of a very thin tube (called a capillary tube), there is no opportunity for the molecules to form into a sphere. But they still try to form a surface with the smallest area. The result is a shape that is dished into the material the water is touching. The effect causes water to be held between the bristles of a brush or within the pores of a sponge or a garment. This is called CAPILLARITY.

The property of capillarity can be used to great advantage, especially since both natural and synthetic materials can be created with capillary pores.

Sponges, paper towels, diapers, and toilet tissue are examples of materials that are designed to have pores that are big and numerous enough to hold the maximum possible amount of water within the material. The material may well have several CRIMPED layers, all designed to provide a large surface area and extra pores (see wetability on pages 28 and 29).

The force of surface tension can produce some remarkable effects. Not only will it draw water into fibrous or POROUS materials, but it will also pull water up a narrow, or capillary, tube.

As the liquid flows upward under the surface tension force, it pulls an increasingly heavy volume of water with it. Eventually the weight of the water balances the surface tension force, and the water stops rising.

Capillarity can have both advantages and disadvantages. Capillary rise makes water move up into dry topsoils and plant root zones from wet subsoils but also from wet soils into the bricks of buildings. That could make walls damp and encourage the growth of fungus. That is why a

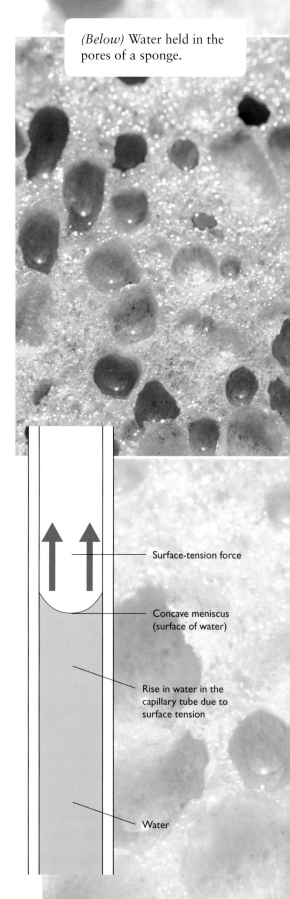

(Below) Water held in the pores of a sponge.

Surface-tension force

Concave meniscus (surface of water)

Rise in water in the capillary tube due to surface tension

Water

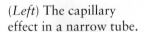

(*Left*) The capillary effect in a narrow tube.

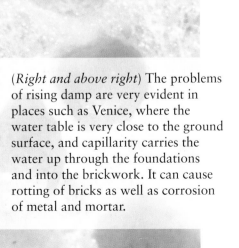

(*Right and above right*) The problems of rising damp are very evident in places such as Venice, where the water table is very close to the ground surface, and capillarity carries the water up through the foundations and into the brickwork. It can cause rotting of bricks as well as corrosion of metal and mortar.

waterproof membrane such as a sheet of polyethylene is inserted within the brickwork of a house just above ground level. It is called a damp-proof layer.

In a closed upright tube where the space above the water is a vacuum, capillarity will cause water to rise up the tube until the weight of water balances atmospheric pressure. As a result, a column of water like this can be used as a BAROMETER. In fact, the height is so great (just under 10 m) that another and heavier liquid—mercury—is used to make barometers a more usable size.

(Above) Damp-proof membrane in a house wall.

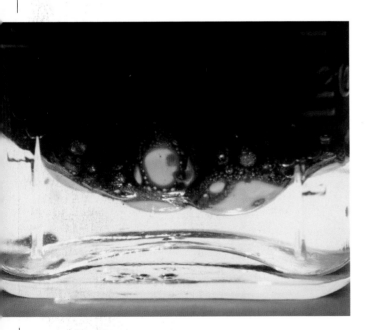

(Above) When oil and water are put together, the water and oil separate because they are immiscible, with the more dense water sinking below the less dense oil.

Immiscibility

The surface-tension effect applies not just to water droplets in air but also to water in a liquid with which it does not mix. This kind of liquid is called IMMISCIBLE. Oil is an example. When water is dripped into oil, it makes globules, which then sink to the bottom of the oil.

Waterproofing and wetting agents

The surface tension effect is so strong in water that it is sometimes difficult to get water to enter a material, especially if it is dry. This effect can be put to good use by making an umbrella, for example, out of a material with very small gaps between the threads. To further resist water and add to the waterproofing effect, the material can be treated with a substance that repels water, such as silicone. Materials such as silicone are HYDROPHOBIC, that is, they repel molecules of water at the molecular level. But they cannot bridge large gaps and only act on surfaces, so a close weave such as on umbrellas coupled with a surface treatment of silicone produces the best results.

The same effects that help with waterproofing can make it very difficult to wet a material thoroughly, for example, a garment that needs washing. To get over this problem, the surface tension of the water needs to be reduced. This is one of the properties of a DETERGENT, for it contains a chemical called a WETTING AGENT that mixes with the water and lowers the attraction of water molecules to each other. Wetting agents of this kind are called SURFACTANTS.

See **Vol. 7: Fibers** to find out more about waterproof material.

(*Below*) Layers of tissue designed to absorb water. This paper has been treated with a chemical surfactant that reduces the surface tension of water and so allows water to flow into the pores of the tissue even more easily.

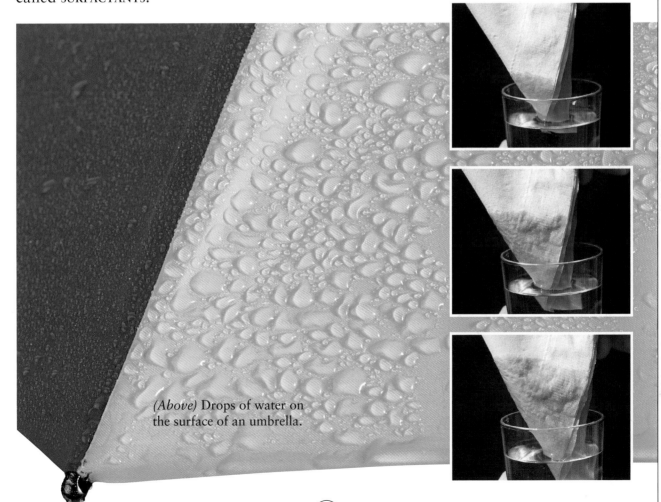

(*Above*) Drops of water on the surface of an umbrella.

Soap and detergents

Two substances are commonly used to help remove greasy materials that would not normally be taken away by water. Soap is a waxy solid produced by reacting sodium hydroxide with fat. When soap is mixed with water, it breaks up into long molecules. One end is attracted to water (it is called HYDROPHILIC), while the other end repels water (it is hydrophobic)—(see also page 45 for more on hydrophilic and hydrophobic substances). This means that the hydrophobic ends are attracted to the surfaces of all objects because that keeps them away from water.

When a soap molecule gets close to a piece of dirt, the hydrophobic ends stick to it. However, to remove the dirt, it has to be rolled around, perhaps by rubbing hands together or by the drum in a washing machine. That gives the soap molecules the chance to stick onto all the surfaces of the dirt, wrapping it up, and thus stopping it from sticking back to the surface it was making dirty. At the same time, the hydrophilic ends of the soap molecules are attracted into the

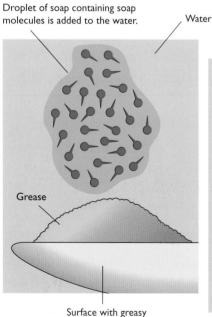

(Above and below)
The action of soap or a detergent.

Droplet of soap containing soap molecules is added to the water.

Water

Grease

Surface with greasy dirt—skin, clothes, and so on

The tails of the soap or detergent molecules are attracted to and embed themselves in the surface of the grease. The water-loving (hydrophilic) heads are left in the water.

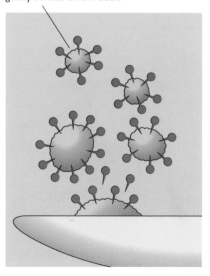

The heads of the molecules are repelled by one another and so, combined with some agitation of the water, will lift into the water with the grease particles. The previously greasy surface is now clean.

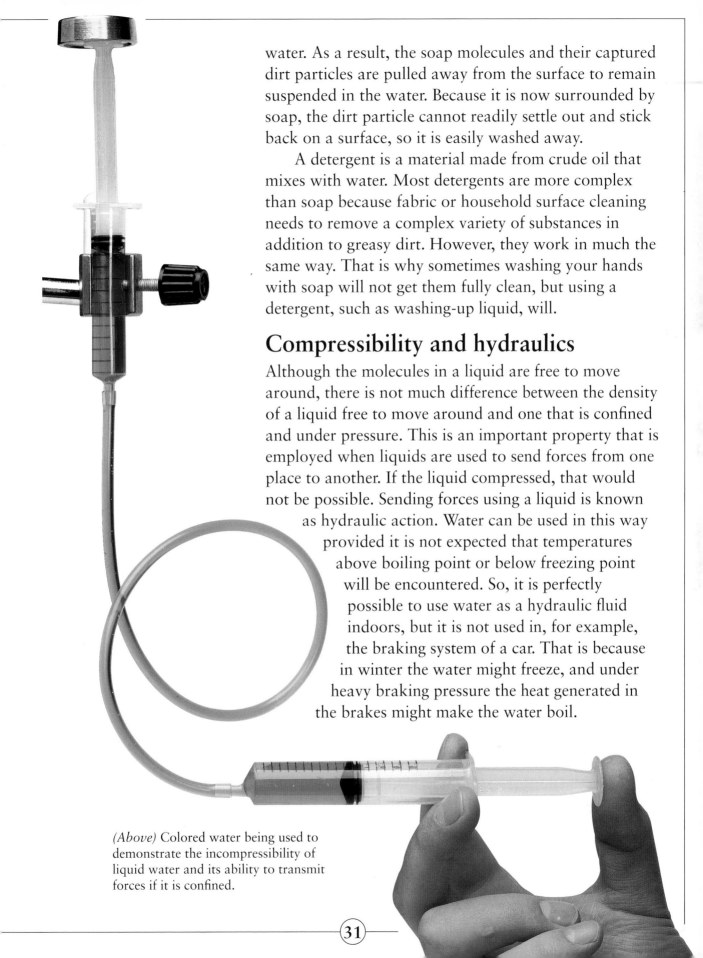

water. As a result, the soap molecules and their captured dirt particles are pulled away from the surface to remain suspended in the water. Because it is now surrounded by soap, the dirt particle cannot readily settle out and stick back on a surface, so it is easily washed away.

A detergent is a material made from crude oil that mixes with water. Most detergents are more complex than soap because fabric or household surface cleaning needs to remove a complex variety of substances in addition to greasy dirt. However, they work in much the same way. That is why sometimes washing your hands with soap will not get them fully clean, but using a detergent, such as washing-up liquid, will.

Compressibility and hydraulics

Although the molecules in a liquid are free to move around, there is not much difference between the density of a liquid free to move around and one that is confined and under pressure. This is an important property that is employed when liquids are used to send forces from one place to another. If the liquid compressed, that would not be possible. Sending forces using a liquid is known as hydraulic action. Water can be used in this way provided it is not expected that temperatures above boiling point or below freezing point will be encountered. So, it is perfectly possible to use water as a hydraulic fluid indoors, but it is not used in, for example, the braking system of a car. That is because in winter the water might freeze, and under heavy braking pressure the heat generated in the brakes might make the water boil.

(Above) Colored water being used to demonstrate the incompressibility of liquid water and its ability to transmit forces if it is confined.

Viscosity

The runniness of a liquid is called its VISCOSITY. You may think that water is a very runny substance, but it is much less runny than some other liquids, for example, gasoline.

In many liquids the warmer the liquid, the runnier it becomes. This shows very dramatically with oil, for example, where cooking oil is far more runny when heated than when poured cold straight from a bottle. However, the difference is not very obvious with water. This is yet another unusual property of water. It means that we can use cold water in much the same way as hot water without having to worry about its viscosity. That would not be true for oil.

Water, however, behaves differently when other substances are added to it, creating a mixture called a SOLUTION. In general, additives make the solution "thicker" and as a consequence "stickier," and the viscosity changes with temperature.

Heat conduction and storage

Water is very good at holding heat. Water is said to have a very high HEAT CAPACITY, or thermal capacity. Water can also conduct heat quite well, certainly very much better than air. That is why, for example, an ocean changes temperature very slowly through the seasons: It takes a long time for the warm, sunlit surface waters to conduct some heat downward. But because water can store heat, the oceans are also slow to cool down in winter, thus keeping many coastal regions warmer than they otherwise would be.

One of the most common domestic uses of the heat capacity of water is in central heating systems. Water is heated in a boiler and then pumped to radiators in various rooms. By using water, a considerable amount of heat can be stored per unit volume, and this keeps down the amount of liquid that has to be pumped around the system.

In some energy-efficient houses there is a large tank of water beneath the basement. During the hot,

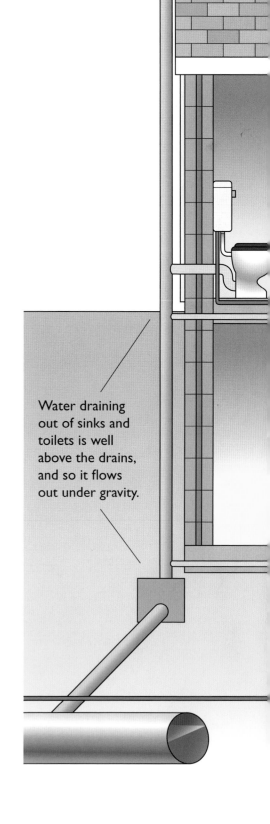

Water draining out of sinks and toilets is well above the drains, and so it flows out under gravity.

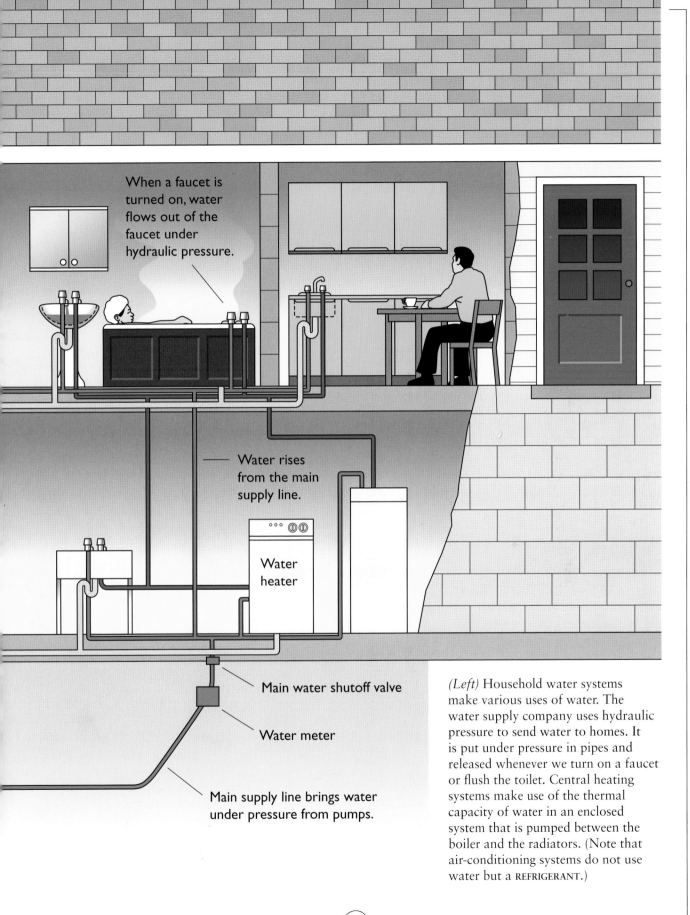

When a faucet is turned on, water flows out of the faucet under hydraulic pressure.

Water rises from the main supply line.

Water heater

Main water shutoff valve

Water meter

Main supply line brings water under pressure from pumps.

(Left) Household water systems make various uses of water. The water supply company uses hydraulic pressure to send water to homes. It is put under pressure in pipes and released whenever we turn on a faucet or flush the toilet. Central heating systems make use of the thermal capacity of water in an enclosed system that is pumped between the boiler and the radiators. (Note that air-conditioning systems do not use water but a REFRIGERANT.)

(*Left*) The high thermal capacity of water is used extensively in cooling systems for power plants. They also use the property that as water changes state from liquid to solid, it absorbs large amounts of heat energy. In the cooling towers the water condenses again, releasing heat to the atmosphere.

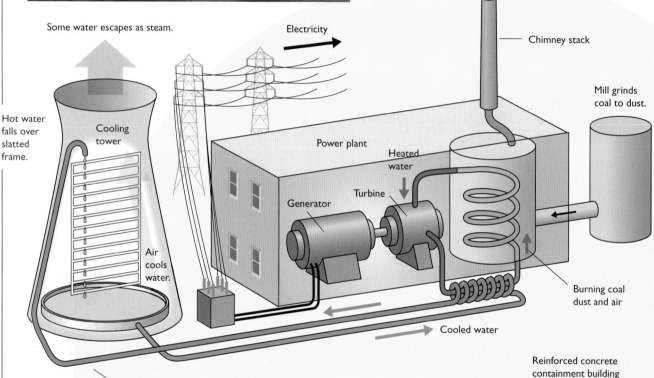

Some water escapes as steam.

Hot water falls over slatted frame.

Cooling tower

Air cools water.

Electricity

Power plant

Generator

Turbine

Heated water

Waste gases

Chimney stack

Mill grinds coal to dust.

Burning coal dust and air

Cooled water

Hot water from power plant

(*Above*) Water is used as a cooling system in thermal power plants.

(*Right*) Water is used for cooling in nuclear power plants. In this case the water has to remain contained for safety reasons.

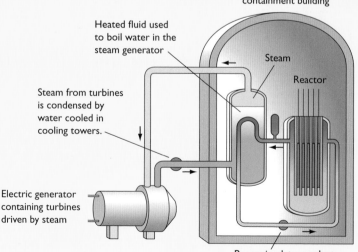

Reinforced concrete containment building

Heated fluid used to boil water in the steam generator

Steam

Reactor

Steam from turbines is condensed by water cooled in cooling towers.

Electric generator containing turbines driven by steam

Pump circulates coolant.

sunny days of summer heat is absorbed by water flowing in roof-mounted tubes, and the hot water is pumped down to the water-storage tank. Heat is exchanged with the water in the tank. During the winter the heated water in this well-insulated storage tank is used as a source of heating.

Hydraulic power generation

Commonly known as hydroelectric power generation, this important use of water relies on the POTENTIAL ENERGY stored by a body of water.

All substances have potential energy due to their position. Since they flow, liquids can be used to transfer this form of energy into KINETIC ENERGY in a controlled way. The kinetic, or movement, energy can be transferred through a turbine to drive a shaft of an electric generator, thus converting the potential energy of water through kinetic energy to electrical energy.

(*Below*) A hydroelectric power plant is an excellent example of how water can be used to convert potential energy to kinetic energy (and thus eventually be the cause of electricity power generation).

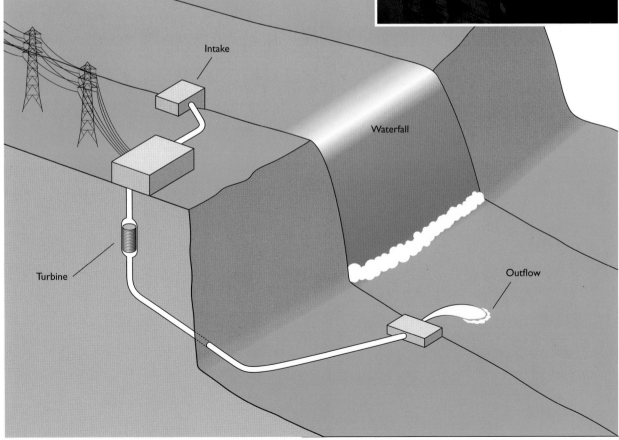

Intake

Waterfall

Turbine

Outflow

Putting out fires

Water is noncombustible, which means that it will not catch fire. When the water is heated, the oxygen will not separate from the hydrogen in the water molecule, and so it will not provide more oxygen to fuel the flames. Instead, the water changes state from liquid to gas, and heat is lost in the air. Because the water is held together with hydrogen bonds, the energy required to vaporize the liquid is particularly high, and so extra heat energy has to be transferred from the burning material to cause vaporization.

At the same time, water, with its high thermal capacity, is very effective at lowering the temperature of a fire, often below the combustion point of the

(Below) Automatic fire extinguishers are set off by an increase in air temperature. Water is jetted out to douse flames before they become difficult to control.

(Right) Fires can be put out with a variety of substances; but water is not only abundant and cheap, it is also particularly effective.

burning material. Furthermore, water smothers the burning substance, preventing oxygen from getting to it and so cutting off the cause of combustion.

Water containing wetting agents to reduce surface tension can also be applied to structures near a fire. This so-called "wet" water, when applied with foaming agents, clings to surfaces better than ordinary water.

Convection of heat

When a solid is heated, the heat moves from one molecule to another. This process, CONDUCTION, was described above. The solid swells a little and become less dense as it warms because it has more energy in it, and the molecules begin to vibrate more vigorously. However, the molecules never change position.

When a liquid such as water is heated, the heated molecules (for example, those touching the bottom of a saucepan on a stove) receive heat by conduction, and they expand and so become less dense, just like a solid. But the molecules in a liquid are not locked together in the same way as in a solid. Instead, they can move, and less dense molecules start to rise up through the colder, more dense molecules like corks bobbing to the surface. This, in turn, causes cold molecules to flow in to take the place of the rising hot molecules. In their turn the cold molecules are heated and begin to rise.

In this way an upward flow of warmer molecules is balanced by a downward flow of colder molecules. This causes a flow of water to begin, which is called CONVECTION.

Convection depends on heat being added to the water from *below*. If the source of heat is near the surface, then convection cannot happen. This has very important results. In nature most ocean water is heated from above by the Sun. As a result, the heated water remains on the surface and does not mix with the cold water below. That is why convection is uncommon in the world's oceans, and why the heat that penetrates into the deeper regions of the oceans is from conduction.

The effects of both conduction and convection are widely used in designs.

(Below) Convection in a saucepan of water heated from below.

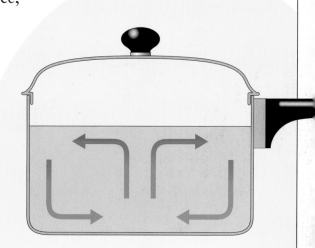

A coffee percolator is an example of where convection is used. The water is heated from below so that convection will carry the water around and around, and the water will get evenly hot.

In a hot water tank, on the other hand, the heating element may be near the top, with cold water entering at the bottom. When a faucet is turned on, the hot water flows out of the top of the tank, pushed out by the inflowing cold. Here the objective is to keep hot water for the faucets at the top of the tank without it mixing with cold water that has just entered at the bottom of the tank. If this happened, the water reaching the faucet would be cooler.

As you can see, by careful manipulation of the properties of convection, quite different results can be achieved.

Buoyancy

When you drop a log onto the surface of water, it floats. For this to happen, the weight (downward force) of the log must be balanced by an equal upward force produced by the water.

The upward force is called BUOYANCY. It was first discovered by the Greek scientist Archimedes, and the principle of buoyancy is called Archimedes' principle.

Archimedes' principle is often thought of as applying to floating objects, but it also applies equally well to submerged objects. Thus a stone that feels heavy on the ground in air will be much lighter in water. This important fact helps explain why ships float, why rivers can carry boulders, as well as why submarines can remain underwater without having to balance their entire weight

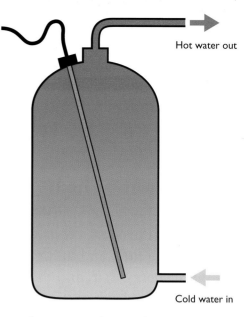

(Above) Water heaters have elements in the top of the tanks so that the upper part of the cylinder gets hot. This reduces convection and prevents the hot water mixing with the cold. Hot water can then be drawn from the top of the tank and cold water introduced at the bottom.

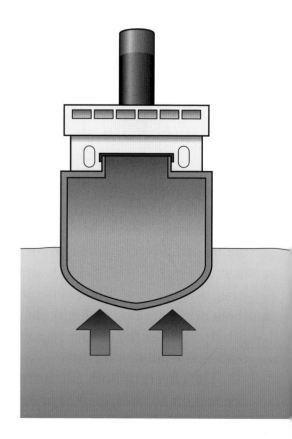

(Right) Buoyancy occurs because the upthrust on the bottom of a floating object balances the weight of the object.

in air, how fish can swim in water, and so on. Water's buoyancy makes it an excellent medium for transporting heavy loads, but its viscosity tends to make this slow.

Sound

Water easily transmits sound energy. When a sound wave moves through water, it causes molecules to move in the direction of the sound. As each molecule moves, it bumps against its neighbor, passing on energy that allows the next molecule to move. Once the extra energy has been passed on, the molecules return to their original places.

This kind of wave depends on the density of the material it moves through. We are used to sound moving quite slowly through air. That is because air is not a very dense material. Sound moves, on average, three times faster through a liquid like water than through air. (It goes even faster through dense solids.)

Diffusion and stirring

The way that molecules of one substance mix by themselves with another is called DIFFUSION. Substances diffuse very slowly in any liquid. This is in great contrast to what happens in air or in any other gas where diffusion is rapid (allowing us to smell things, for example).

As a result, when you make a cup of coffee, you normally add the coffee and sugar and then stir because, if you didn't, the coffee and the sugar would be very slow to mix with the water.

Because diffusion is not a helpful process in liquids, stirring is widely used to mix them. However, stirring does not have to be by giant spoons or rotors. It can also be achieved by forcing jets of air under high pressure through the liquid. This is used, for example, in sewage plants and in factories that need to separate metals from powdered rock.

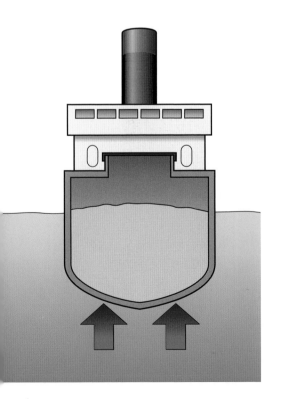

(*Left*) This carrot shows osmosis at work in nature. A healthy carrot has been put in distilled water (*far left*) and a strong salt solution (*left*). Osmosis insures a transfer of water out of the cells in the carrot that has been put in the salt solution, making it shrivel. Water, and its transfer between cells are extremely important for living things. About 95% of this carrot is water.

Osmosis

Molecules move around in a solution until they are evenly spread among one another. This is DIFFUSION. But sometimes solutions are separated by SEMIPERMEABLE MEMBRANES, which are thin sheets of material with holes just big enough to allow water molecules to go through but not big enough to allow the flow of molecules of dissolved substances.

When a weak SOLUTION is placed next to a concentrated solution and separated by a semipermeable membrane, something curious happens. Water starts to flow from the dilute to the concentrated solution. That is because substances are only at rest when their concentrations are the same everywhere.

Take the example of a concentrated salt solution

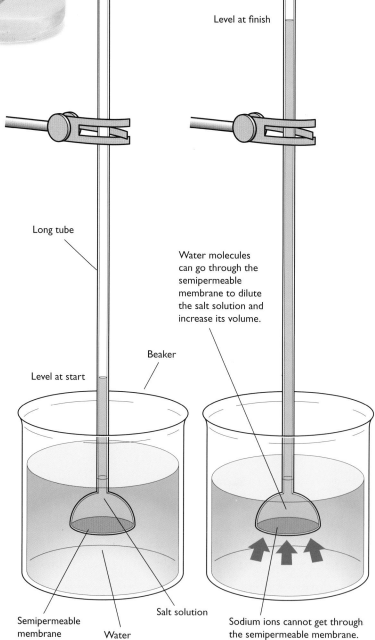

Level at finish

Long tube

Water molecules can go through the semipermeable membrane to dilute the salt solution and increase its volume.

Beaker

Level at start

Semipermeable membrane

Water

Salt solution

Sodium ions cannot get through the semipermeable membrane.

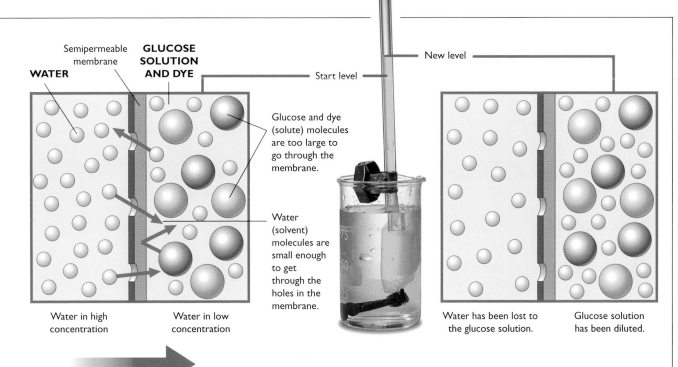

WATER   Semipermeable membrane   **GLUCOSE SOLUTION AND DYE**

Start level   New level

Glucose and dye (solute) molecules are too large to go through the membrane.

Water (solvent) molecules are small enough to get through the holes in the membrane.

Water in high concentration   Water in low concentration

Water has been lost to the glucose solution.   Glucose solution has been diluted.

The net diffusion of water molecules is down the concentration gradient and into the glucose solution.

(Above) In this experiment a piece of special viscose tubing has been used to demonstrate osmosis. Glucose solution mixed with a dye was placed in the tube and lowered into a beaker of distilled water. It took only a few minutes for the level of the liquid in the tube to rise by about 1 cm. The water molecules have moved through the membrane from the beaker, where they are in high concentration, and into the glucose solution, where they are in low concentration in order to address this imbalance.

divided from a more dilute solution by a semipermeable membrane. The sodium and chloride ions in the salt cannot get through the membrane. However, the water can move from the dilute side (where the water is in greatest concentration) to the concentrated side in order to balance the concentrations. However, in doing this, the water builds up in what was the concentrated side.

The height of water that can be supported by this process is quite considerable. It is used by plants to get water into their roots and up to the leaves. It is also used in animals, including us, for example, to extract food from the gut.

Osmosis can work in industry too. It is used, for example, to concentrate fruit juice without boiling the juice (by placing the natural, weak juice next to a concentrated salt solution), for desalting seawater, and for helping clean up sewage.

The reverse osmosis process is a membrane process used to extract fresh water from brackish inland water, the salt content of which, though undesirable, is considerably below that of seawater. Brine, subjected to pressure, is forced against a semipermeable membrane; fresh water passes through, while the concentrated mineral salts remain behind. The membranes are made from suitable polymer materials such as cellulose acetate.

The reverse osmosis process is not as commonly used as the distillation desalination processes available—see page 50.

4: The universal solvent

A solution is produced when a solid, liquid, or gas becomes mixed in a liquid in such a way that it breaks up into molecules or ions. We then say that the substance has dissolved in the liquid. It is important to remember that substances dissolved (solutes) in the liquid (solvent) do not have themselves to be liquids. Blood and seawater, for example, both contain, among many other substances, oxygen gas (thus allowing animals to breathe).

It is also important to remember that a solution is a mixture. The substances inside the solution have not reacted together and changed into new substances. Instead, the dissolved substances have merely been broken up into their molecules or ions and

(Below) Because salt is in solution, it can be extracted from the sea by the simple process of evaporation. The salty seawater is stored in ponds, and the water allowed to evaporate. What is left behind is a mixture of salts that can then be used with food or processed in a chemical plant to separate them. This is an important method of getting common salt (sodium chloride) as well as other salts.

(Below) Salt water contains a wide range of substances dissolved in it. They become apparent if the water is allowed to evaporate. The white encrustation that remains is a mixture of all of the salts that were dissolved in the water. Other dissolved substances, such as oxygen, escape into the air during evaporation.

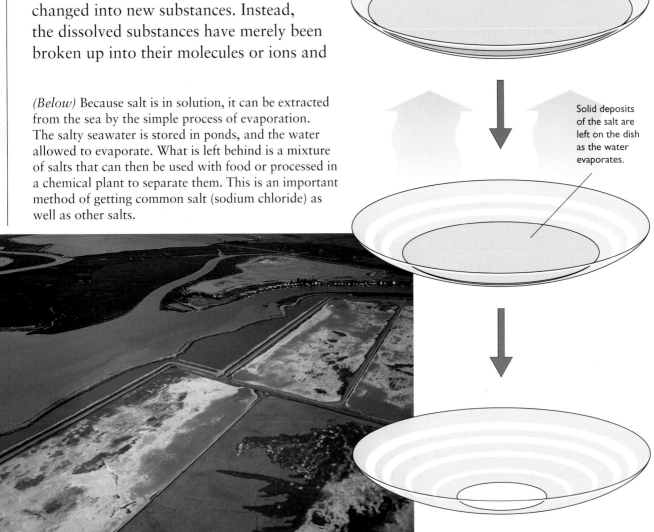

Salt water

Solid deposits of the salt are left on the dish as the water evaporates.

the molecules or ions scattered evenly among the molecules of the liquid it has dissolved in. We call a solution in water an AQUEOUS SOLUTION (aqueous meaning "of water").

Why water is not quite a "universal solvent"

One of the most noticeable things about water is how well it can dissolve other substances and therefore how important it is to us and to all living things. That is why it has been called the "universal solvent."

When you stir coffee and sugar into hot water, you produce a hot aqueous solution. The biggest aqueous solution on Earth is, of course, the world's oceans, with their vast amount of dissolved salts. These salts are, by the way, one of the world's greatest reservoirs of many chemical substances.

Water also forms the basis of all living things. It makes up 70% by weight of the human body. Water is able to transport substances in solution, both other liquids, such as foods and ions such as sodium and potassium, as well as gases, such as oxygen and carbon dioxide.

As we have seen earlier in this book, water is a *polar solvent*. As such, it has the special property of being able to dissolve many ionic solids such as sodium chloride—salt. This produces a solution in water.

Water makes ionic substances dissolve in the following way. When an ionic substance like salt (which is a chemical made of sodium and chlorine) is dissolved in water, the positive parts of the salt (the sodium ions) are attracted to the negative end of the water molecule, while the negative parts of the salts (the chlorine ions) are attracted to the positive parts of the water molecule. Because the parts of the salt are dragged to opposite parts of the water molecules, the salt splits apart and so dissolves.

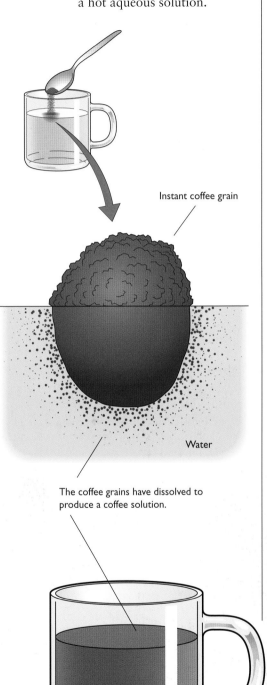

(Below) When you stir coffee and sugar into hot water, you produce a hot aqueous solution.

Instant coffee grain

Water

The coffee grains have dissolved to produce a coffee solution.

(*Right*) Sodium chloride, or salt, is an ionic solid made of sodium and chloride ions. It is crystalline and has a lattice structure.

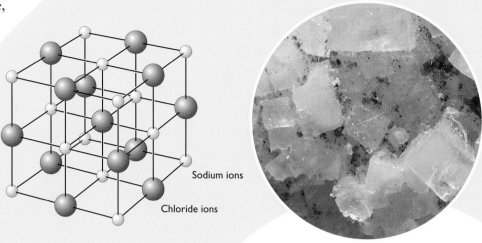

Sodium ions

Chloride ions

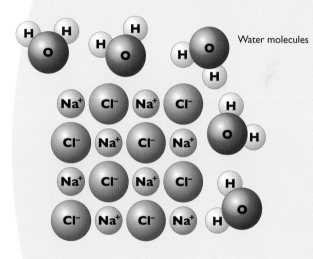

Water molecules

(*Left*) The ionic solid is put into water. The ions attract the water molecules, and the charged ends of the water molecules are attracted to oppositely charged ions.

(*Below*) Free ions go into solution surrounded by water molecules. This is now described as an aqueous solution of sodium chloride. It will conduct electricity.

As ions break away with the water molecules, the lattice starts to split up.

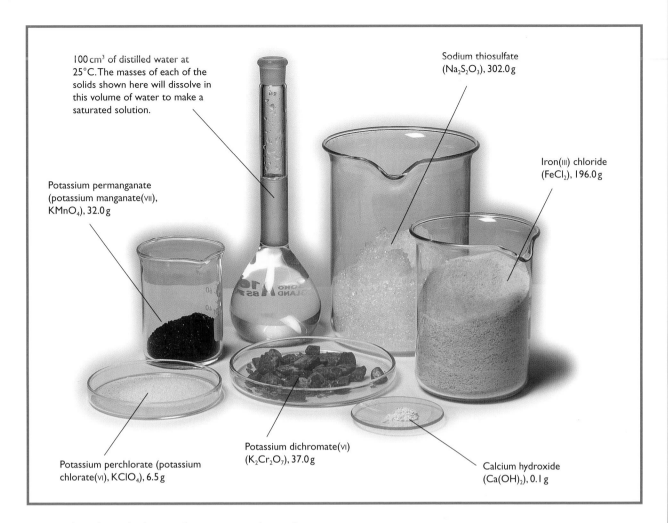

100 cm³ of distilled water at 25°C. The masses of each of the solids shown here will dissolve in this volume of water to make a saturated solution.

Sodium thiosulfate ($Na_2S_2O_3$), 302.0 g

Iron(III) chloride ($FeCl_2$), 196.0 g

Potassium permanganate (potassium manganate(VII), $KMnO_4$), 32.0 g

Potassium perchlorate (potassium chlorate(VI), $KClO_4$), 6.5 g

Potassium dichromate(VI) ($K_2Cr_2O_7$), 37.0 g

Calcium hydroxide ($Ca(OH)_2$), 0.1 g

How hydrophilic substances dissolve

Not all substances that dissolve in water are like salt (that is, they are not made up of ionic particles that can be split apart). Alcohol (as found in alcoholic beverages, and whose scientific name is ethanol) dissolves readily in water. Most of an alcoholic drink is, in fact, water. In this case the alcohol molecules are similar to water molecules (although much more complicated) and so can attract water molecules as well as attracting themselves.

Why hydrophobic substances do not dissolve

Why will water not dissolve everything? The answer to this is that "like dissolves like." The substance dissolving in water has to have something in common with water in the way it is made. Fat is unlike water in its structure and behavior, and so it does

How much will dissolve?

Throughout this book you have seen how water easily dissolves other substances. But how much can be dissolved? Is it the same for every substance? To find out, it is possible to add more and more of a chemical to water until no more will dissolve. You know no more will dissolve because no matter how much the solution is stirred, some solid remains at the bottom.

Different substances have widely varying solubilities.

The temperature of the solution also has an important effect on the amount of a substance that will dissolve. In general, substances become much more soluble as the temperature is raised.

not dissolve. This is, of course, important in our bodies too, because otherwise everything we are made of would dissolve! Washing-up detergent and soap, on the other hand, have properties that are similar to water and some that are similar to fat. So, they are able to act as a chemical go-between and allow water to be used to remove fat.

Water at high temperature and pressure

Interestingly, water only works on the "like dissolves like" principle when it is at normal temperatures. If water is made very hot (and kept liquid by keeping it under pressure), it will dissolve even substances it is not similar to. This is an extremely odd but useful property. It is used in industry, for example, for destroying organic wastes. In this process oxygen is mixed with the organic material under very hot high pressure. The organic material combusts (catches fire) with a flame burning under water, the very opposite to the way water normally behaves when it is cold and not under pressure!

Conducting electricity

Pure water cannot conduct electricity. That is because water molecules carry no electrical charge. But water is rarely pure and usually contains substantial quantities of materials.

Dissolved solids such as salt are called ionic substances. That is, they can split up into negatively charged and positively charged particles, which are called IONS. Although ionic solids such as salt do not conduct electricity, and pure water is also a poor conductor, when salt dissolves in water, the ions are free to move around. When it dissociates in this way, the salt is said to have been IONIZED. As a result, the aqueous solution of salt in water is a very good conductor of electricity. Solutions like this are called ELECTROLYTES.

Not all solids dissolve and form ions, however. When sugar dissolves, it forms molecules with no charge. Sugary water is therefore as good an insulator as pure water.

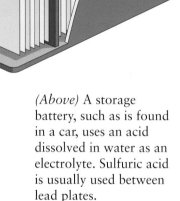

(Above) A storage battery, such as is found in a car, uses an acid dissolved in water as an electrolyte. Sulfuric acid is usually used between lead plates.

Industrial separation of a salt solution using electrolysis

Water is heated and pumped down into salt rocks. They dissolve to produce a concentrated sodium chloride solution called brine. Using water means that large quantities of salt can be dissolved, and the resulting solution can be broken down by electrolysis. The solution will conduct electricity and behave as an electrolyte because of the free ions. The electric current will take the ions out of solution, forming hydrogen and chlorine gases and leaving a concentrated sodium hydroxide solution. This is an industrial method for producing these various chemicals and is known as the Castner-Kellner process.

Seawater contains so much dissolved salt (sodium chloride) that you can taste it. However, most mixtures contain such small amounts that we are unaware of them, and they only show up when they give water special properties, as we shall see below.

Ions are the equivalent of the moving electrons in a wire. It is the movement of the ions through the water that makes electricity flow.

Once a substance has become ionized in water, the conductivity of the water increases dramatically. Thus it is only DISTILLED WATER that is a good insulator; normal faucet water contains many dissolved salts and so is a relatively good conductor of electricity. That is why it is so dangerous to touch electrical equipment with wet hands.

Salt and many strong acids quickly break up, or dissociate, in water and carry electrical charges. That is one reason, for example, that a car battery contains sulfuric acid. But not all substances do this. For example, weak acids such as citric acid only break up to a small extent. As a result, citric acid is just a reasonable, but not a good, conductor of electricity.

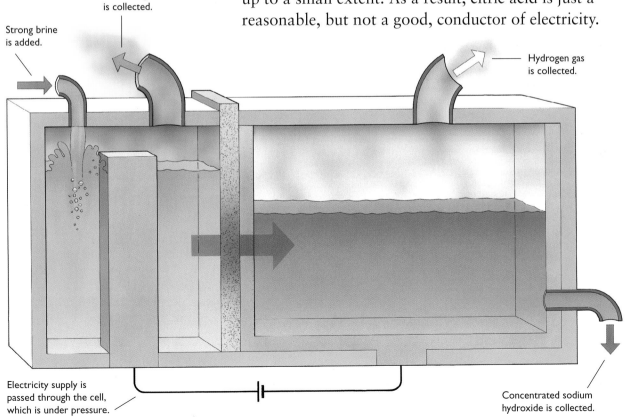

Chlorine gas is collected.

Strong brine is added.

Hydrogen gas is collected.

Electricity supply is passed through the cell, which is under pressure.

Concentrated sodium hydroxide is collected.

When carbon dioxide dissolves in water, it produces another weak acid—carbonic acid—but again it has only fair conducting properties.

The properties of dissociation can be used to great effect, for example, in the industrial separation of salt from salt water (see pages 50 to 51).

Introducing medicines to the blood

Blood (which doctors regard as an electrolyte) is water containing a huge array of other substances, one of which (hemoglobin) colors the water red. But there are many others, and they are all important in carrying vital nutrients and other substances to the cells of our bodies.

One important result of this is when people are ill and need extra supplies of vital substances. They are introduced into the blood using an intravenous drip, that is, the water containing the substances needed to help life are fed in through a needle directly into the blood supply.

Pills are the same. They contain substances that dissolve and so enter the blood supply, where they can be carried to the cells that need the chemicals the pills contain.

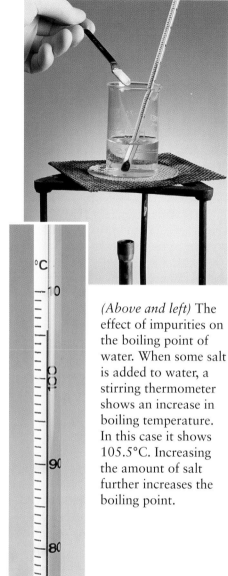

(Above and left) The effect of impurities on the boiling point of water. When some salt is added to water, a stirring thermometer shows an increase in boiling temperature. In this case it shows 105.5°C. Increasing the amount of salt further increases the boiling point.

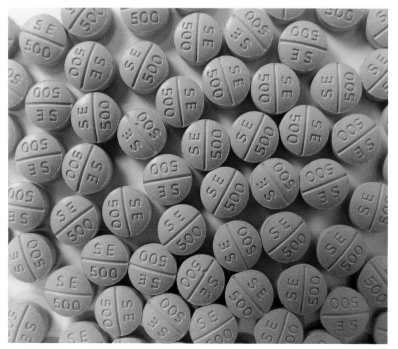

(Left) Because the body is essentially a hugely complex chemical factory whose transportation system is water, the method for getting nourishment and drugs (medicines) to the body quickly in times of illness is to take them orally (when they dissolve in the stomach) or intravenously (when they enter the bloodstream directly).

You will find an apparatus for making distilled water in most laboratories. It is an example of simple distillation, designed to remove the naturally dissolved substances from the water.

Faucet water is boiled, and the water vapor (steam) produced is sent over a large surface area of coiled tubing that is cooled with cold faucet water. The water vapor condenses and is drained off and collected as distilled water.

Pure liquids and mixed liquids

Pure water contains only water molecules made from a combination of oxygen and hydrogen atoms (with the formula H_2O). This is the water that we can get by the process called DISTILLATION, in which we boil water and condense and collect the water vapor.

Pure liquids behave in a very precise way. We can figure out the temperature at which they will boil and freeze or melt. So, for example, we can say with certainty that the boiling point of pure water is 100°C, and its freezing or melting point is 0°C (at the normal pressure of the air at sea level). But if we add impurities to it and make the liquid a mixture (solution or suspension), then both the boiling and freezing points change—and not in any easily predictable way.

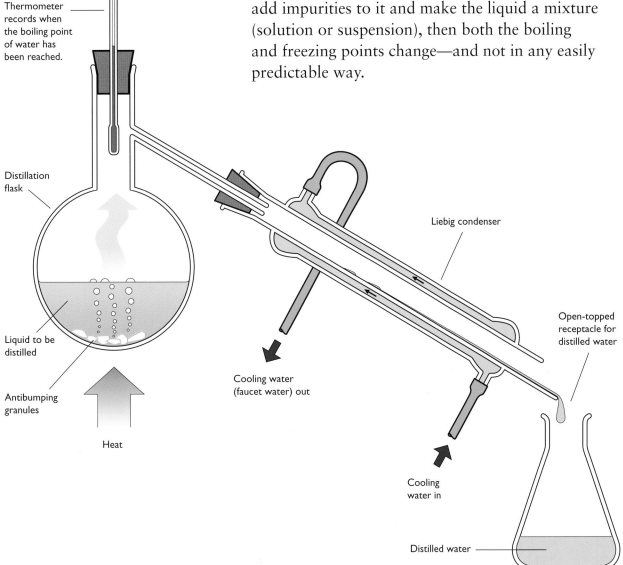

Thermometer records when the boiling point of water has been reached.

Distillation flask

Liquid to be distilled

Antibumping granules

Heat

Cooling water (faucet water) out

Cooling water in

Liebig condenser

Open-topped receptacle for distilled water

Distilled water

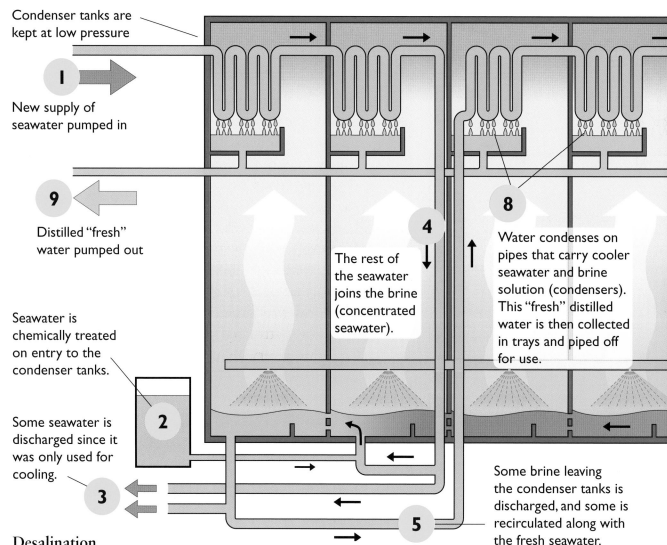

Condenser tanks are kept at low pressure

1 New supply of seawater pumped in

9 Distilled "fresh" water pumped out

Seawater is chemically treated on entry to the condenser tanks.

Some seawater is discharged since it was only used for cooling.

2

3

4 The rest of the seawater joins the brine (concentrated seawater).

8 Water condenses on pipes that carry cooler seawater and brine solution (condensers). This "fresh" distilled water is then collected in trays and piped off for use.

5

Some brine leaving the condenser tanks is discharged, and some is recirculated along with the fresh seawater.

Desalination

Distillation of water is used on a very large scale in countries where water is scarce, for example, in desert countries. In this case it is known as desalination.

Water is distilled from salt water by spraying the seawater into heated tanks in which the air pressure is low. This causes water to boil and steam to form at a temperature below its normal boiling point. This process is called "flashing." The steam is then condensed onto cooled collector tubes. The process is called flash distillation. It uses a very large amount of energy and can only work where energy is plentiful, such as in some Middle Eastern states. The largest desalination plants are on the Arabian Peninsula, and here a series of flash chambers is used to improve the efficiency of the process.

Large plants can produce millions of cubic meters of water a day.

We can say that in general, the more that a solid is dissolved in water, the more concentrated the solution and the lower the freezing point will be. This property is used, for example, when salt is put on roads to reduce icing. The freezing point of a high concentration of salt in water is several degrees below zero (the exact value depending on the concentration of the salt). We also find that salty water boils above 100°C. In fact, water used for boiling vegetables with salt added for flavoring has the unintended effect of raising the boiling point of the water and so slightly speeding up the cooking.

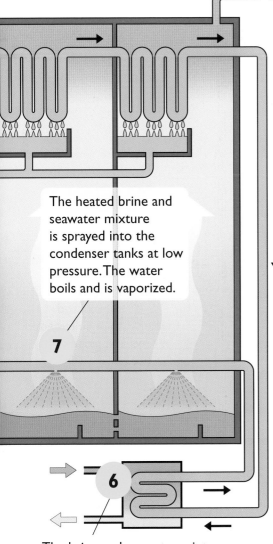

The heated brine and seawater mixture is sprayed into the condenser tanks at low pressure. The water boils and is vaporized.

7

6

The brine and seawater mixture are heated using a steam heat exchanger.

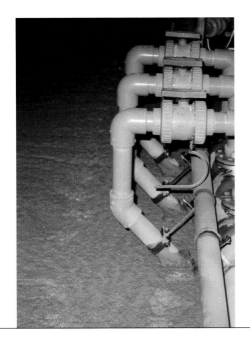

Surface removal of dissolved substances

You might think that the particles dissolved in a liquid are evenly scattered, and in many examples this is true. But it is not always the case. Some dissolved substances concentrate at the surface of the liquid.

One case in which this is useful is when treating polluted water. Soluble organic materials (such as sewage) are found concentrated at the surface of the water more than at depth. However, they are not floating on the water; they are simply attracted to the water surface.

This is very convenient because it allows water-treatment plants to get rid of much pollution by a process called FROTH SEPARATION.

Jets of air are directed through the polluted water, forming air bubbles. The organic materials concentrate on the bubble surfaces. The bubbles then rise to make a froth on the surface of the water. A large arm then skims off the froth and with it much of the sewage.

Metal-extraction plants use the same process to separate small particles of metal from powdered rock waste—a process known as froth flotation.

Another way to use this property is to add a small amount of a liquid such as a special type of alcohol to reservoir water. Alcohol completely dissolves in water; it does not float on water as oil would. However, the alcohol concentrates at the surface and forms a film just a few molecules thick over the water surface. This is not enough to prevent air from getting into the water, but it

(*Left*) In froth separation air is bubbled through the water, and some things stick to the surfaces of the bubbles. They can then be scraped off.

does keep many water molecules from evaporating from the surface. This technique is used in places with hot climates, such as Arizona, where loss of water from reservoirs can be a major problem.

Water reactions that give out heat

It is quite common for heat to be given up or given out when two substances mix. This is called an EXOTHERMIC REACTION. One common example of this is when caustic soda (sodium hydroxide, NaOH) is added to water. Enormous amounts of heat are given out, which makes the water very hot.

The purpose of dissolving caustic soda in water is to unblock drains by dissolving grease and fat. Grease and fat will not dissolve in water at any temperature. Grease and fat will not easily dissolve in cold caustic soda either. But grease and fat will melt when they are hot. So, by increasing the temperature by adding caustic soda to water, the caustic soda can dramatically increase the amount of grease it is able to dissolve.

Water reactions with metals

Many substances dissolve in water. But that is not a chemical reaction. If the water is evaporated, the dissolved material re-forms. But water is able to react with some substances and so make completely new substances. In some cases water can behave just like a weak ACID. In other cases it can behave as a weak ALKALI.

When a very reactive metal such as potassium is put in water, it reacts violently. The water reacts with the potassium and gives off hydrogen gas and a great deal of heat energy. The result is that the hydrogen gas catches fire.

The hydrogen gas is produced when water molecules lose one hydrogen atom each. What remains is one hydrogen atom and one oxygen atom, which immediately react with potassium to form the new substance, potassium hydroxide.

(*Above*) When water is added to calcium oxide, a violently exothermic reaction occurs. You can see the steam condensing on the side of the beaker above.

The oxide blocks swell, and calcium hydroxide is formed (*below*).

(Above) A piece of potassium burning on the surface of water.

(Below) Corrosion of iron requires the presence of both water and air. If an iron nail is placed in a small tube that is partly filled with distilled water and left open to the air, the part of the nail in the water will gradually corrode.

Although very reactive metals react in this violent way, less reactive metals also react. Most metals react in water containing dissolved oxygen. During the chemical reaction that takes place, the metal is oxidized and a surface coating produced. We call this CORROSION.

The water and the air dissolved in it act as an electrical conducting solution. One of the metals most severely affected by this reaction is iron. The particular form of corrosion that occurs is called RUST. It is also a chemical reaction.

However, not all metals react as quickly. There is, in fact an order of reactivity (see chart on page 54). Magnesium is more reactive in water than iron, for example. The importance of this is that if a piece of magnesium is wrapped around a nail and the two placed in water, then the magnesium reacts with the water, and the iron remains uncorroded. This is an extremely important result because it means that, for example, if blocks of a reactive metal are placed on the hull of a ship, it is the blocks of metal that corrode, while the hull remains undamaged. Such protection saves huge

Distilled water in open container so that air can mix freely

Rust develops on nail within 24 hours.

(Right) However, if some similar new, clean nails are put in a DESICCATOR, an apparatus designed to keep the air completely dry, and if yet more nails are put in a tube that is then filled with freshly prepared distilled water (which has been boiled in preparation and so contains no dissolved oxygen), little corrosion is seen even after many weeks in either case.

Element
Potassium (K)
Sodium (Na)
Calcium (Ca)
Magnesium (Mg)
Aluminum (Al)
Manganese Mn)
Chromium (Cr)
Zinc (Zn)
Iron (Fe)
Cadmium (Cd)
Tin (Sn)
Lead (Pb)
Hydrogen (H$_2$)
Copper (Cu)
Mercury (Hg)
Silver (Ag)
Platinum (Pt)
Gold (Au)

amounts in repair bills to ships and other outdoor equipment. It is even used on some cars.

It is important to realize that this only works because the iron is less reactive than the other metal. If the iron is placed in contact with a less reactive metal, then the iron corrodes very fast. In this example tin and iron are placed together, and the tin remains unaffected. Tin is often used as a protective plating over the iron because it is unreactive; but should the tin plating be damaged, the iron will quickly corrode.

If the water is heated above boiling point (superheated), it will react directly with metals (even when no air is present), and hydrogen gas is given off. Boilers that produce superheated water are therefore liable to become corroded.

In fact, the only metals that do not react with water (either at ordinary temperatures or when

(Above) The reactivity series of metals.

(Below) An iron nail with strips of magnesium (*left*) and tin (*right*) attached.

Water containing phenolphthalein indicator shows where electrochemical reactions are taking place.

Iron nail remains uncorroded.

Magnesium strip is corroding within a few minutes.

Over about a week the iron nail is corroded.

Tin strip remains uncorroded.

(Below) The precipitation of calcium carbonate on heating elements and pipes in heating systems causes the buildup of scale. Scale has a low heat conductivity and so stops the heating elements from working. It may also cause overheating of the element. This is, in fact, one of the principal causes of failure in electrical water elements. The problem can be reduced by the use of water softeners. They work by reacting with the calcium and magnesium salts, producing substances that remain soluble. An alternative method is to remove some of the offending materials using special clay filters.

superheated) are the metals known as the noble metals—for example, gold, silver, and platinum.

Hard water

One of the results of water being such a good solvent is that it has many substances dissolved in it. In particular, water commonly contains calcium and magnesium compounds, mostly as bicarbonates.

The presence of calcium and magnesium compounds makes the water less likely to produce a lather when soap is used. When it becomes difficult to get a lather, usually when the water contains more than 120 milligrams of calcium and magnesium salts in each liter of water, the water is described as hard. It is common to find hard water associated with river water and groundwater supplies in areas of limestone or chalk rock.

If the water hardness is mainly from calcium bicarbonate, then the water is called TEMPORARILY HARD WATER. If the water is boiled, the bicarbonate changes to carbonate and is precipitated out as, for example, the fur on a kettle element. This is a permanent chemical reaction.

All other compounds of calcium and magnesium produce PERMANENT HARDNESS because boiling does not bring them out of solution. Instead, they prevent soap (but not detergents) from producing a lather.

Breaking up water molecules into hydrogen and oxygen

Water will not readily dissociate, or break up, into hydrogen and oxygen gases unless a large amount of energy is applied, as, for example, when a large electric current is sent through it. This explains why water does not break up into gases when heated, but instead changes from liquid to gas.

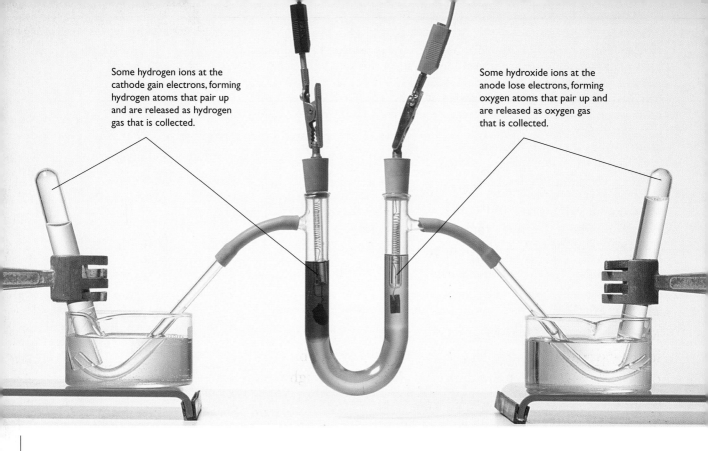

Some hydrogen ions at the cathode gain electrons, forming hydrogen atoms that pair up and are released as hydrogen gas that is collected.

Some hydroxide ions at the anode lose electrons, forming oxygen atoms that pair up and are released as oxygen gas that is collected.

Decomposing water using electrolysis

Two ends of a U-tube contain electrodes that will introduce electricity into the water. When the power is switched on, gas is immediately formed as a bubbling stream at both electrodes.

By looking at the tubes collecting the gas, it is possible to see that twice the volume of gas is produced in one tube as in the other.

This demonstration also verifies the formula for water (H_2O) because twice as much hydrogen as oxygen is released.

The process of turning water into oxygen and hydrogen using electricity is called ELECTROLYSIS.

Hydration

Many substances have a strong affinity for water, and they will soak it up from the air. That is why in a humid atmosphere paper, cookies, and many other substances go limp, for example. But it is also true of chemicals. Concentrated sulfuric acid, for example, has a strong affinity for water and will take it from the atmosphere. This can be shown by placing a jar of water and a jar of concentrated sulfuric acid in a larger jar and then shutting the lid. Over a few weeks the water disappears, and the sulfuric acid appears to increase in volume. The sulfuric acid has sucked water from the air, allowing more evaporation to take place from the water. The acid may look the same color as at the start, but the change in volume tells you it is more dilute.

Some chemicals change color when they absorb water, or HYDRATE. Blue cobalt chloride and white anhydrous copper sulfate are two of them.

(Right) Rainwater is weakly acidic. As it percolates through limestone rock, it reacts to form soluble calcium bicarbonate. When this solution reaches a cave, some of the carbon dioxide escapes from the water, and the calcium bicarbonate changes to insoluble calcium carbonate, which is precipitated. In this picture a cave straw about 5 mm across is forming. It is the starting point for a stalactite.

Some materials also give out great amounts of heat when they absorb water. When water is added to lime (calcium oxide), it heats up dramatically. At the same time, the calcium oxide blocks swell and crack apart. Within a few minutes the hard lime block has been changed into a soft white powder. This is how lime is produced for use as a fertilizer on fields.

Acid water

Many gases dissolve in water. This is especially important in the case of water droplets in the air because the billions of droplets produce such a large surface area through which gases can dissolve.

The most common gas in the air, which will dissolve in water and create an acid, is carbon dioxide. It reacts with the water to produce carbonic acid (H_2CO_3). This is the acid that is mainly responsible for sculpturing limestone landscapes throughout the world.

Stronger acids can be produced by the reaction of industrial pollutant gases and water. Nitrogen dioxide reacts to produce nitrous acid, and sulfur dioxide reacts to produce sulfurous acid, which in turn change to nitric and sulfuric acid. These are the gases that produce acid rain. Acid rain is primarily responsible for the deterioration of buildings containing limestone, for example, churches and cathedrals.

The reaction of these gases with water can be shown in a simple demonstration. If an indicator solution is added to water, it first shows a blue color, indicating lack of acidity. However, as gas is bubbled through the water, the color of the indicator changes until the color is orange, showing acid conditions.

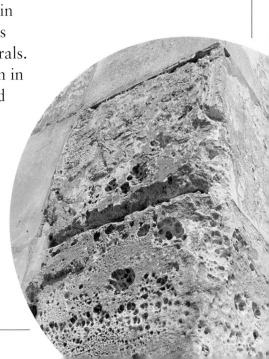

(Right) Where air pollutants increase the acidity of rainwater, the corrosion of certain stones such as limestone speeds up. Damage due to acid rain can weaken buildings structurally, such as is seen here.

Set Glossary

ACID RAIN: Rain that falls after having been contaminated by acid gases produced by power plants, vehicle exhausts, and other man-made sources.

ACIDITY: The tendency of a liquid to behave like an acid, reacting with metals and alkalis.

ADDITION POLYMERIZATION: The building blocks of many plastics (or polymers) are simple molecules called monomers. Monomers can be converted into polymers by making the monomers link to one another to form long chains in head-to-tail fashion. This is called addition polymerization or chain polymerization. It is most often used to link vinyl monomers to produce, for example, PVC, or polyvinyl chloride polymer.
See also **CONDENSATION POLYMERIZATION**

ADHESIVE: Any substance that can hold materials together simply by using some kind of surface attachment. In some cases this is a chemical reaction; in other cases it is a physical attraction between molecules of the adhesive and molecules of the substance it sticks to.

ADOBE: Simple unbaked brick made with mud, straw, and dung. It is dried in the open air. In this form it is very vulnerable to the effects of rainfall and so is most often found in desert areas or alternatively is protected by some waterproof covering, for example, thatch, straw, or reeds.

ALKALI: A base, or substance that can neutralize acids. In glassmaking an alkali is usually potassium carbonate and used as a flux to lower the melting point of the silica.

ALKYD: Any kind of synthetic resin used for protective coatings such as paint.

ALLOY: A metal mixture made up of two or more elements. Most of the elements used to make an alloy are metals. For example, brass is an alloy of copper and zinc, but carbon is an exception and used to make steel from iron.

AMALGAM: An alloy of mercury and one or more other metals. Dentist's filling amalgam traditionally contains mercury, silver, and tin.

AMPHIBIOUS: Adapted to function on both water and land.

AMORPHOUS: Shapeless and having no crystalline form. Glass is an amorphous solid.

ANION: An ion with a negative charge.

ANNEALING: A way of making a metal, alloy, or glass less brittle and more easy to work (more ductile) by heating it to a certain temperature (depending on the metal), holding it at that temperature for a certain time, and then cooling to room temperature.

ANODIZING: A method of plating metal by electrically depositing an oxide film onto the surface of a metal. The main purpose is to reduce corrosion.

ANTICYCLONE: A region of the Earth's atmosphere where the pressure is greater than average.

AQUEOUS SOLUTION: A substance dissolved in water.

ARTIFACT: An object of a previous time that was created by humans.

ARTIFICIAL DYE: A dye made from a chemical reaction that does not occur in nature. Dyes made from petroleum products are artificial dyes.

ARTIFICIAL FIBER: A fiber made from a material that has been manufactured, and that does not occur naturally. Rayon is an example of an artificial fiber.
Compare to **SYNTHETIC**

ATMOSPHERE: The envelope of gases that surrounds the Earth.

ATOM: The smallest particle of an element; a nucleus and its surrounding electrons.

AZO: A chemical compound that contains two nitrogen atoms joined by a double bond and each linked to a carbon atom. Azon compounds make up more than half of all dyes.

BARK: The exterior protective sheath of the stem and root of a woody plant such as a tree or a shrub. Everything beyond the cambium layer.

BAROMETER: An instrument for measuring atmospheric pressure.

BASE METAL: Having a low value and poorer properties than some other metals. Used, for example, when describing coins that contain metals other than gold or silver.

BAST FIBERS: A strong woody fiber that comes from the phloem of plants and is used for rope and similar products. Flax is an example of a bast fiber.

BATCH: A mixture of raw materials or products that are processes in a tank or kiln. This process produces small amounts of material or products and can be contrasted to continuous processes. Batch processing is used to make metals, alloys, glass, plastics, bricks, and other ceramics, dyes, and adhesives.

BAUXITE: A hydrated impure oxide of aluminum. It is the main ore used to obtain aluminum metal. The reddish-brown color of bauxite is caused by impurities of iron oxides.

BINDER: A substance used to make sure the pigment in a paint sticks to the surface it is applied to.

BIOCERAMICS: Ceramic materials that are used for medical and dental purposes, mainly as implants and replacements.

BLAST FURNACE: A tall furnace charged with a mixture of iron ore, coke, and limestone and used for the refining (smelting) of iron ore. The name comes from the strong blast of air used during smelting.

BLOWING: Forming a glass object by blowing into a gob of molten glass to form a bubble on the end of a blowpipe.

BOLL: The part of the cotton seed that contains the cotton fiber.

BOILING POINT: The temperature at which a liquid changes to a vapor. Boiling points change with atmospheric pressure.

BOND: A transfer or a sharing of electrons by two or more atoms. There are a number of kinds of chemical bonds, some very strong, such as covalent bonding and ionic bonding, and others quite weak, as in hydrogen bonding. Chemical bonds form because the linked molecules are more stable than the unlinked atoms from which they are formed.

BOYLE'S LAW: At constant temperature and for a given mass of gas the volume of the gas is inversely proportional to the pressure that builds up.

BRITTLE: Something that has almost no plasticity and so shatters rather than bends when a force is applied.

BULL'S EYE: A piece of glass with concentric rings marking the place where the blowpipe was attached to the glass. It is the central part of a pane of crown glass.

BUOYANCY: The tendency of an object to float if it is less dense than the liquid it is placed in.

BURN: A combustion reaction in which a flame is produced. A flame occurs where gases combust and release heat and light. At least two gases are therefore required if there is to be a flame.

CALORIFIC: Relating to the production of heat.

CAMBIUM: A thin growing layer that separates the xylem and phloem in most plants, and that produces new cell layers.

CAPACITOR: An electronic device designed for the temporary storage of electricity.

CAPILLARY ACTION, CAPILLARITY: The process by which surface tension forces can draw a liquid up a fine-bore tube.

CARBOHYDRATES: One of the main constituents of green plants, containing compounds of carbon, hydrogen, and oxygen. The main kinds of carbohydrate are sugars, starches, and celluloses.

CARBON COMPOUNDS: Any compound that includes the element carbon. Carbon compounds are also called organic compounds because they form an essential part of all living organisms.

CARBON CYCLE: The continuous movement of carbon between living things, the soil, the atmosphere, oceans, and rocks, especially those containing coal and petroleum.

CAST: To pour a liquid metal, glass, or other material into a mold and allow it to cool so that it solidifies and takes on the shape of the mold.

CATALYST: A substance that speeds up a chemical reaction but itself remains unchanged. For example, platinum is used in a catalytic converter of gases in the exhausts leaving motor vehicles.

CATALYTIC EFFECT: The way a substance helps speed up a reaction even though that substance does not form part of the reaction.

CATHODIC PROTECTION: The technique of protecting a metal object by connecting it to a more easily oxidizable material. The metal object being protected is made into the cathode of a cell. For example, iron can be protected by coupling it with magnesium.

CATION: An ion with a positive charge, often a metal.

CELL: A vessel containing two electrodes and a liquid substance that conducts electricity (an electrolyte).

CELLULOSE: A form of carbohydrate. *See* **CARBOHYDRATE**

CEMENT: A mixture of alumina, silica, lime, iron oxide, and magnesium oxide that is burned together in a kiln and then made into a powder. It is used as the main ingredient of mortar and as the adhesive in concrete.

CERAMIC: A crystalline nonmetal. In a more everyday sense it is a material based on clay that has been heated so that it has chemically hardened.

CHARRING: To burn partly so that some of a material turns to carbon and turns black.

CHINA: A shortened version of the original "Chinese porcelain," it also refers to various porcelain objects such as plates and vases meant for domestic use.

CHINA CLAY: The mineral kaolinite, which is a very white clay used as the basis of porcelain manufacture.

CLAY MINERALS: The minerals, such as kaolinite, illite, and montmorillonite, that occur naturally in soils and some rocks, and that are all minute platelike crystals.

COKE: A form of coal that has been roasted in the absence of air to remove much of the liquid and gas content.

COLORANTS: Any substance that adds a color to a material. The pigments in paints and the chemicals that make dyes are colorants.

COLORFAST: A dye that will not "run" in water or change color when it is exposed to sunlight.

COMPOSITE MATERIALS: Materials such as plywood that are normally regarded as a single material, but that themselves are made up of a number of different materials bonded together.

COMPOUND: A chemical consisting of two or more elements chemically bonded together, for example, calcium carbonate.

COMPRESSED AIR: Air that has been squashed to reduce its volume.

COMPRESSION: To be squashed.

COMPRESSION MOLDING: The shaping of an object, such as a headlight lens, which is achieved by squashing it into a mold.

CONCRETE: A mixture of cement and a coarse material such as sand and small stones.

CONDENSATION: The process of changing a gas to a liquid.

CONDENSATION POLYMERIZATION: The production of a polymer formed by a chain of reactions in which a water molecule is eliminated as every link of the polymer is formed. Polyester is an example.

CONDUCTION: (i) The exchange of heat (heat conduction) by contact with another object, or (ii) allowing the flow of electrons (electrical conduction).

CONDUCTIVITY: The property of allowing the flow of heat or electricity.

CONDUCTOR: (i) Heat—a material that allows heat to flow in and out of it easily. (ii) Electricity—a material that allows electrons to flow through it easily.

CONTACT ADHESIVE: An adhesive that, when placed on the surface to be joined, sticks as soon as the surfaces are placed firmly together.

CONVECTION: The circulating movement of molecules in a liquid or gas as a result of heating it from below.

CORRODE/CORROSION: A reaction usually between a metal and an acid or alkali in which the metal decomposes. The word is used in the sense of the metal being eaten away and dangerously thinned.

CORROSIVE: Causing corrosion, that is, the oxidation of a metal. For example, sodium hydroxide is corrosive.

COVALENT BONDING: The most common type of strong chemical bond, which occurs when two atoms share electrons. For example, oxygen O_2.

CRANKSHAFT: A rodlike piece of a machine designed to change linear into rotational motion or vice versa.

CRIMP: To cause to become wavy.

CRUCIBLE: A ceramic-lined container for holding molten metal, glass, and so on.

CRUDE OIL: A chemical mixture of petroleum liquids. Crude oil forms the raw material for an oil refinery.

CRYSTAL: A substance that has grown freely so that it can develop external faces.

CRYSTALLINE: A solid in which the atoms, ions, or molecules are organized into an orderly pattern without distinct crystal faces.

CURING: The process of allowing a chemical change to occur simply by waiting a while. Curing is often a process of reaction with water or with air.

CYLINDER GLASS: An old method of making window glass by blowing a large bubble of glass, then swinging it until it forms a cylinder. The ends of the cylinder are then cut off with shears and the sides of the cylinder allowed to open out until they form a flat sheet.

DECIDUOUS: A plant that sheds its leaves seasonally.

DECOMPOSE: To rot. Decomposing plant matter releases nutrients back to the soil and in this way provides nourishment for a new generation of living things.

DENSITY: The mass per unit volume (for example, g/c^3).

DESICCATE: To dry up thoroughly.

DETERGENT: A cleaning agent that is able to turn oils and dirts into an emulsion and then hold them in suspension so they can be washed away.

DIE: A tool for giving metal a required shape either by striking the object with the die or by forcing the object over or through the die.

DIFFUSION: The slow mixing of one substance with another until the two substances are evenly mixed. Mixing occurs because of differences in concentration within the mixture. Diffusion works rapidly with gases, very slowly with liquids.

DILUTE: To add more of a solvent to a solution.

DISSOCIATE: To break up. When a compound dissociates, its molecules break up into separate ions.

DISSOLVED: To break down a substance in a solution without causing a reaction.

DISTILLATION: The process of separating mixtures by condensing the vapors through cooling. The simplest form of distillation uses a Liebig condenser arranged with just a slight slope down to the collecting vessel. When the liquid mixture is heated and vapors are produced, they enter the water cooled condenser and then flow down the tube, where they can be collected.

DISTILLED WATER: Water that has its dissolved solids removed by the process of distillation.

DOPING: Adding an impurity to the surface of a substance in order to change its properties.

DORMANT: A period of inactivity such as during winter, when plants stop growing.

DRAWING: The process in which a piece of metal is pulled over a former or through dies.

DRY-CLEANED: A method of cleaning fabrics with nonwater-based organic solvents such as carbon tetrachloride.

DUCTILE: Capable of being drawn out or hammered thin.

DYE: A colored substance that will stick to another substance so that both appear to be colored.

EARLY WOOD: The wood growth put on the spring of each year.

EARTHENWARE: Pottery that has not been fired to the point where some of the clay crystals begin to melt and fuse together and is thus slightly porous and coarser than stoneware or porcelain.

ELASTIC: The ability of an object to regain its original shape after it has been deformed.

ELASTIC CHANGE: To change shape elastically.

ELASTICITY: The property of a substance that causes it to return to its original shape after it has been deformed in some way.

ELASTIC LIMIT: The largest force that a material can stand before it changes shape permanently.

ELECTRODE: A conductor that forms one terminal of a cell.

ELECTROLYSIS: An electrical-chemical process that uses an electric current to cause the breakup of a compound and the movement of metal ions in a solution. It is commonly used in industry for purifying (refining) metals or for plating metal objects with a fine, even metal coat.

ELECTROLYTE: An ionic solution that conducts electricity.

ELECTROMAGNET: A temporary magnet that is produced when a current of electricity passes through a coil of wire.

ELECTRON: A tiny, negatively charged particle that is part of an atom. The flow of electrons through a solid material such as a wire produces an electric current.

ELEMENT: A substance that cannot be decomposed into simpler substances by chemical means, for example, silver and copper.

EMULSION: Tiny droplets of one substance dispersed in another.

EMULSION PAINT: A paint made of an emulsion that is water soluble (also called latex paint).

ENAMEL: A substance made of finely powdered glass colored with a metallic oxide and suspended in oil so that it can be applied with a brush. The enamel is then heated, the oil burns away, and the glass fuses. Also used colloquially to refer to certain kinds of resin-based paint that have extremely durable properties.

ENGINEERED WOOD PRODUCTS: Wood products such as plywood sheeting made from a combination of wood sheets, chips or sawdust, and resin.

EVAPORATION: The change of state of a liquid to a gas. Evaporation happens below the boiling point.

EXOTHERMIC REACTION: A chemical reaction that gives out heat.

EXTRUSION: To push a substance through an opening so as to change its shape.

FABRIC: A material made by weaving threads into a network, often just referred to as cloth.

FELTED: Wool that has been hammered in the presence of heat and moisture to change its texture and mat the fibers.

FERRITE: A magnetic substance made of ferric oxide combined with manganese, nickel, or zinc oxide.

FIBER: A long thread.

FILAMENT: (i) The coiled wire used inside a light bulb. It consists of a high-resistance metal such as tungsten that also has a high melting point. (ii) A continuous thread produced during the manufacture of fibers.

FILLER: A material introduced in order to give bulk to a substance. Fillers are used in making paper and also in the manufacture of paints and some adhesives.

FILTRATE: The liquid that has passed through a filter.

FLOOD: When rivers spill over their banks and cover the surrounding land with water.

FLUID: Able to flow either as a liquid or a gas.

FLUORESCENT: A substance that gives out visible light when struck by invisible waves, such as ultraviolet rays.

FLUX: A substance that lowers the melting temperature of another substance. Fluxes are use in glassmaking and in melting alloys. A flux is used, for example, with a solder.

FORMER: An object used to control the shape or size of a product being made, for example, glass.

FOAM: A material that is sufficiently gelatinous to be able to contain bubbles of gas. The gas bulks up the substances, making it behave as though it were semirigid.

FORGE: To hammer a piece of heated metal until it changes to the desired shape.

FRACTION: A group of similar components of a mixture. In the petroleum industry the light fractions of crude oil are those with the smallest molecules, while the medium and heavy fractions have larger molecules.

FRACTIONAL DISTILLATION: The separation of the components of a liquid mixture by heating them to their boiling points.

FREEZING POINT: The temperature at which a substance undergoes a phase change from a liquid to a solid. It is the same temperature as the melting point.

FRIT: Partly fused materials of which glass is made.

FROTH SEPARATION: A process in which air bubbles are blown through a suspension, causing a froth of bubbles to collect on the surface. The materials that are attracted to the bubbles can then be removed with the froth.

FURNACE: An enclosed fire designed to produce a very high degree of heat for melting glass or metal or for reheating objects so they can be further processed.

FUSING: The process of melting particles of a material so they form a continuous sheet or solid object. Enamel is bonded to the surface of glass this way. Powder-formed metal is also fused into a solid piece. Powder paints are fused to the surface by heating.

GALVANIZING: The application of a surface coating of zinc to iron or steel.

GAS: A form of matter in which the molecules take no definite shape and are free to move around to uniformly fill any vessel they are put in. A gas can easily be compressed into a much smaller volume.

GIANT MOLECULES: Molecules that have been formed by polymerization.

GLASS: A homogeneous, often transparent material with a random noncrystalline molecular structure. It is achieved by cooling a molten substance very rapidly so that it cannot crystallize.

GLASS CERAMIC: A ceramic that is not entirely crystalline.

GLASSY STATE: A solid in which the molecules are arranged randomly rather than being formed into crystals.

GLOBAL WARMING: The progressive increase in the average temperature of the Earth's atmosphere, most probably in large part due to burning fossil fuels.

GLUE: An adhesive made from boiled animal bones.

GOB: A piece of near-molten glass used by glass-blowers and in machines to make hollow glass vessels.

GRAIN: (i) The distinctive pattern of fibers in wood. (ii) Small particles of a solid, including a single crystal.

GRAPHITE: A form of the element carbon with a sheetlike structure.

GRAVITY: The attractive force produced because of the mass of an object.

GREENHOUSE EFFECT: An increase in the global air temperature as a result of heat released from burning fossil fuels being absorbed by carbon dioxide in the atmosphere.

GREENHOUSE GAS: Any of various gases that contribute to the greenhouse effect, such as carbon dioxide.

GROUNDWATER: Water that flows naturally through rocks as part of the water cycle.

GUM: Any natural adhesive of plant origin that consists of colloidal polysaccharide substances that are gelatinous when moist but harden on drying.

HARDWOOD: The wood from a nonconiferous tree.

HEARTWOOD: The old, hard, nonliving central wood of trees.

HEAT: The energy that is transferred when a substance is at a different temperature than that of its surroundings.

HEAT CAPACITY: The ratio of the heat supplied to a substance compared with the rise in temperature that is produced.

HOLOGRAM: A three-dimensional image reproduced from a split laser beam.

HYDRATION: The process of absorption of water by a substance. In some cases hydration makes a substance change color, but in all cases there is a change in volume.

HYDROCARBON: A compound in which only hydrogen and carbon atoms are present. Most fuels are hydrocarbons, for example, methane.

HYDROFLUORIC ACID: An extremely corrosive acid that attacks silicate minerals such as glass. It is used to etch decoration onto glass and also to produce some forms of polished surface.

HYDROGEN BOND: A type of attractive force that holds one molecule to another. It is one of the weaker forms of intermolecular attractive force.

HYDROLYSIS: A reversible process of decomposition of a substance in water.

HYDROPHILIC: Attracted to water.

HYDROPHOBIC: Repelled by water.

IMMISCIBLE: Will not mix with another substance, for example, oil and water.

IMPURITIES: Any substances that are found in small quantities, and that are not meant to be in the solution or mixture.

INCANDESCENT: Glowing with heat, for example, a tungsten filament in a light bulb.

INDUSTRIAL REVOLUTION: The time, which began in the 18th century and continued through into the 19th century, when materials began to be made with the use of power machines and mass production.

INERT: A material that does not react chemically.

INORGANIC: A substance that does not contain the element carbon (and usually hydrogen), for example, sodium chloride.

INSOLUBLE: A substance that will not dissolve, for example, gold in water.

INSULATOR: A material that does not conduct electricity.

ION: An atom or group of atoms that has gained or lost one or more electrons and so developed an electrical charge.

IONIC BONDING: The form of bonding that occurs between two ions when the ions have opposite charges, for example, sodium ions bond with chloride ions to make sodium chloride. Ionic bonds are strong except in the presence of a solvent.

IONIZE: To change into ions.

ISOTOPE: An atom that has the same number of protons in its nucleus, but that has a different mass, for example, carbon 12 and carbon 14.

KAOLINITE: A form of clay mineral found concentrated as china clay. It is the result of the decomposition of the mineral feldspar.

KILN: An oven used to heat materials. Kilns at quite low temperatures are used to dry wood and at higher temperatures to bake bricks and to fuse enamel onto the surfaces of other substances. They are a form of furnace.

KINETIC ENERGY: The energy due to movement. When a ball is thrown, it has kinetic energy.

KNOT: The changed pattern in rings in wood due to the former presence of a branch.

LAMINATE: An engineered wood product consisting of several wood layers bonded by a resin. Also applies to strips of paper stuck together with resins to make such things as "formica" worktops.

LATE WOOD: Wood produced during the summer part of the growing season.

LATENT HEAT: The amount of heat that is absorbed or released during the process of changing state between gas, liquid, or solid. For example, heat is absorbed when liquid changes to gas. Heat is given out again as the gas condenses back to a liquid.

LATEX: A general term for a colloidal suspension of rubber-type material in water. Originally for the milky white liquid emulsion found in the Para rubber tree, but also now any manufactured water emulsion containing synthetic rubber or plastic.

LATEX PAINT: A water emulsion of a synthetic rubber or plastic used as paint. *See* **EMULSION PAINT**

LATHE: A tool consisting of a rotating spindle and cutters that is designed to produce shaped objects that are symmetrical about the axis of rotation.

LATTICE: A regular geometric arrangement of objects in space.

LEHR: The oven used for annealing glassware. It is usually a very long tunnel through which glass passes on a conveyor belt.

LIGHTFAST: A colorant that does not fade when exposed to sunlight.

LIGNIN: A form of hard cellulose that forms the walls of cells.

LIQUID: A form of matter that has a fixed volume but no fixed shape.

LUMBER: Timber that has been dressed for use in building or carpentry and consists of planed planks.

MALLEABLE: Capable of being hammered or rolled into a new shape without fracturing due to brittleness.

MANOMETER: A device for measuring liquid or gas pressure.

MASS: The amount of matter in an object. In common use the word weight is used instead (incorrectly) to mean mass.

MATERIAL: Anything made of matter.

MATTED: Another word for felted. *See* **FELTED**

MATTER: Anything that has mass and takes up space.

MELT: The liquid glass produced when a batch of raw materials melts. Also used to describe molten metal.

MELTING POINT: The temperature at which a substance changes state from a solid phase to a liquid phase. It is the same as the freezing point.

METAL: A class of elements that is a good conductor of electricity and heat, has a metallic luster, is malleable and ductile, and is formed as cations held together by a sea of electrons. A metal may also be an alloy of these elements and carbon.

METAL FATIGUE: The gradual weakening of a metal by constant bending until a crack develops.

MINERAL: A solid substance made of just one element or compound, for example, calcite minerals contain only calcium carbonate.

MISCIBLE: Capable of being mixed.

MIXTURE: A material that can be separated into two or more substances using physical means, for example, air.

MOLD: A containing shape made of wood, metal, or sand into which molten glass or metal is poured. In metalworking it produces a casting. In glassmaking the glass is often blown rather than poured when making, for example, light bulbs.

MOLECULE: A group of two or more atoms held together by chemical bonds.

MONOMER: A small molecule and building block for larger chain molecules or polymers (mono means "one" and mer means "part").

MORDANT: A chemical that is attracted to a dye and also to the surface that is to be dyed.

MOSAIC: A decorated surface made from a large number of small colored pieces of glass, natural stone, or ceramic that are cemented together.

NATIVE METAL: A pure form of a metal not combined as a compound. Native metals are more common in nonreactive elements such as gold than reactive ones such as calcium.

NATURAL DYES: Dyes made from plants without any chemical alteration, for example, indigo.

NATURAL FIBERS: Fibers obtained from plants or animals, for example, flax and wool.

NEUTRON: A particle inside the nucleus of an atom that is neutral and has no charge.

NOBLE GASES: The members of group 8 of the periodic table of the elements: helium, neon, argon, krypton, xenon, radon. These gases are almost entirely unreactive.

NONMETAL: A brittle substance that does not conduct electricity, for example, sulfur or nitrogen.

OIL-BASED PAINTS: Paints that are not based on water as a vehicle. Traditional artists' oil paint uses linseed oil as a vehicle.

OPAQUE: A substance through which light cannot pass.

ORE: A rock containing enough of a useful substance to make mining it worthwhile, for example, bauxite, the ore of aluminum.

ORGANIC: A substance that contains carbon and usually hydrogen. The carbonates are usually excluded.

OXIDE: A compound that includes oxygen and one other element, for example, Cu_2O, copper oxide.

OXIDIZE, OXIDIZING AGENT: A reaction that occurs when a substance combines with oxygen or a reaction in which an atom, ion, or molecule loses electrons to another substance (and in this more general case does not have to take up oxygen).

OZONE: A form of oxygen whose molecules contain three atoms of oxygen. Ozone high in the atmosphere blocks harmful ultraviolet rays from the Sun, but at ground level it is an irritant gas when breathed in and so is regarded as a form of pollution. The ozone layer is the uppermost part of the stratosphere.

PAINT: A coating that has both decorative and protective properties, and that consists of a pigment suspended in a vehicle, or binder, made of a resin dissolved in a solvent. It dries to give a tough film.

PARTIAL PRESSURE: The pressure a gas in a mixture would exert if it alone occupied the flask. For example, oxygen makes up about a fifth of the atmosphere. Its partial pressure is therefore about a fifth of normal atmospheric pressure.

PASTE: A thick suspension of a solid in a liquid.

PATINA: A surface coating that develops on metals and protects them from further corrosion, for example, the green coating of copper carbonate that forms on copper statues.

PERIODIC TABLE: A chart organizing elements by atomic number and chemical properties into groups and periods.

PERMANENT HARDNESS: Hardness in the water that cannot be removed by boiling.

PETROCHEMICAL: Any of a large group of manufactured chemicals (not fuels) that come from petroleum and natural gas. It is usually taken to include similar products that can be made from coal and plants.

PETROLEUM: A natural mixture of a range of gases, liquids, and solids derived from the decomposed remains of animals and plants.

PHASE: A particular state of matter. A substance can exist as a solid, liquid, or gas and may change between these phases with the addition or removal of energy, usually in the form of heat.

PHOSPHOR: A material that glows when energized by ultraviolet or electron beams, such as in fluorescent tubes and cathode ray tubes.

PHOTOCHEMICAL SMOG: A mixture of tiny particles of dust and soot combined with a brown haze caused by the reaction of colorless nitric oxide from vehicle exhausts and oxygen of the air to form brown nitrogen dioxide.

PHOTOCHROMIC GLASSES: Glasses designed to change color with the intensity of light. They use the property that certain substances, for example, silver halide, can change color (and change chemically) in light. For example, when silver chromide is dispersed in the glass melt, sunlight decomposes the silver halide to release silver (and so darken the lens). But the halogen cannot escape; and when the light is removed, the halogen recombines with the silver to turn back to colorless silver halide.

PHOTOSYNTHESIS: The natural process that happens in green plants whereby the energy from light is used to help turn gases, water, and minerals into tissue and energy.

PIEZOELECTRICS: Materials that produce electric currents when they are deformed, or vice versa.

PIGMENT: Insoluble particles of coloring material.

PITH: The central strand of spongy tissue found in the stems of most plants.

PLASTIC: Material—a carbon-based substance consisting of long chains or networks (polymers) of simple molecules. The word plastic is commonly used only for synthetic polymers. Property—a material is plastic if it can be made to change shape easily and then remain in this new shape (contrast with elasticity and brittleness).

PLASTIC CHANGE: A permanent change in shape that happens without breaking.

PLASTICIZER: A chemical added to rubbers and resins to make it easier for them to be deformed and molded. Plasticizers are also added to cement to make it more easily worked when used as a mortar.

PLATE GLASS: Rolled, ground, and polished sheet glass.

PLIABLE: Supple enough to be repeatedly bent without fracturing.

PLYWOOD: An engineered wood laminate consisting of sheets of wood bonded with resin. Each sheet of wood has the grain at right angles to the one above and below. This imparts stability to the product.

PNEUMATIC DEVICE: Any device that works with air pressure.

POLAR: Something that has a partial electric charge.

POLYAMIDES: A compound that contains more than one amide group, for example, nylon.

POLYMER: A compound that is made of long chains or branching networks by combining molecules called monomers as repeating units. Poly means "many," mer means "part."

PORCELAIN: A hard, fine-grained, and translucent white ceramic that is made of china clay and is fired to a high temperature. Varieties include china.

PORES: Spaces between particles that are small enough to hold water by capillary action, but large enough to allow water to enter.

POROUS: A material that has small cavities in it, known as pores. These pores may or may not be joined. As a result, porous materials may or may not allow a liquid or gas to pass through them. Popularly, porous is used to mean permeable, the kind of porosity in which the pores are joined, and liquids or gases can flow.

POROUS CERAMICS: Ceramics that have not been fired at temperatures high enough to cause the clays to fuse and so prevent the slow movement of water.

POTENTIAL ENERGY: Energy due to the position of an object. Water in a reservoir has potential energy because it is stored up, and when released, it moves down to a lower level.

POWDER COATING: The application of a pigment in powder form without the use of a solvent.

POWDER FORMING: A process of using a powder to fill a mold and then heating the powder to make it fuse into a solid.

PRECIPITATE: A solid substance formed as a result of a chemical reaction between two liquids or gases.

PRESSURE: The force per unit area measured in SI units in Pascals and also more generally in atmospheres.

PRIMARY COLORS: A set of colors from which all others can be made. In transmitted light they are red, blue, and green.

PROTEIN: Substances in plants and animals that include nitrogen.

PROTON: A positively charged particle in the nucleus of an atom that balances out the charge of the surrounding electrons.

QUENCH: To put into water in order to cool rapidly.

RADIATION: The transmission of energy from one body to another without any contribution from the intervening space. *Contrast with* **CONVECTION** and **CONDUCTION**

RADIOACTIVE: A substance that spontaneously emits energetic particles.

RARE EARTHS: Any of a group of metal oxides that are found widely throughout the Earth's rocks, but in low concentrations. They are mainly made up of the elements of the lanthanide series of the periodic table of the elements.

RAW MATERIAL: A substance that has not been prepared, but that has an intended use in manufacturing.

RAY: Narrow beam of light.

RAYON: An artificial fiber made from natural cellulose.

REACTION (CHEMICAL): The recombination of two substances using parts of each substance.

REACTIVE: A substance that easily reacts with many other substances.

RECYCLE: To take once used materials and make them available for reuse.

REDUCTION, REDUCING AGENT: The removal of oxygen from or the addition of hydrogen to a compound.

REFINING: Separating a mixture into the simpler substances of which it is made, especially petrochemical refining.

REFRACTION: The bending of a ray of light as it passes between substances of different refractive index (light-bending properties).

REFRACTORY: Relating to the use of a ceramic material, especially a brick, in high-temperature conditions of, for example, a furnace.

REFRIGERANT: A substance that, on changing between a liquid and a gas, can absorb large amounts of (latent) heat from its surroundings.

REGENERATED FIBERS: Fibers that have been dissolved in a solution and then recovered from the solution in a different form.

REINFORCED FIBER: A fiber that is mixed with a resin, for example, glass-reinforced fiber.

RESIN: A semisolid natural material that is made of plant secretions and often yellow-brown in color. Also synthetic materials with the same type of properties. Synthetic resins have taken over almost completely from natural resins and are available as thermoplastic resins and thermosetting resins.

RESPIRATION: The process of taking in oxygen and releasing carbon dioxide in animals and the reverse in plants.

RIVET: A small rod of metal that is inserted into two holes in metal sheets and then burred over at both ends in order to stick the sheets together.

ROCK: A naturally hard inorganic material composed of mineral particles or crystals.

ROLLING: The process in which metal is rolled into plates and bars.

ROSIN: A brittle form of resin used in varnishes.

RUST: The product of the corrosion of iron and steel in the presence of air and water.

SALT: Generally thought of as sodium chloride, common salt; however, more generally a salt is a compound involving a metal. There are therefore many "salts" in water in addition to sodium chloride.

SAPWOOD: The outer, living layers of the tree, which includes cells for the transportation of water and minerals between roots and leaves.

SATURATED: A state in which a liquid can hold no more of a substance dissolved in it.

SEALANTS: A material designed to stop water or other liquids from penetrating into a surface or between surfaces. Most sealants are adhesives.

SEMICONDUCTOR: A crystalline solid that has an electrical conductivity part way between a conductor and an insulator. This material can be altered by doping to control an electric current. Semiconductors are the basis of transistors, integrated circuits, and other modern electronic solid-state devices.

SEMIPERMEABLE MEMBRANE: A thin material that acts as a fine sieve or filter, allowing small molecules to pass, but holding back large molecules.

SEPARATING COLUMN: A tall glass tube containing a porous disk near the base and filled with a substance such as aluminum oxide that can absorb materials on its surface. When a mixture passes through the columns, fractions are retarded by differing amounts so that each fraction is washed through the column in sequence.

SEPARATING FUNNEL: A pear-shaped glass funnel designed to permit the separation of immiscible liquids by simply pouring off the more dense liquid from the bottom of the funnel, while leaving the less dense liquid in the funnel.

SHAKES: A defect in wood produced by the wood tissue separating, usually parallel to the rings.

SHEEN: A lustrous, shiny surface on a yarn. It is produced by the finishing process or may be a natural part of the yarn.

SHEET-METAL FORMING: The process of rolling out metal into sheet.

SILICA: Silicon dioxide, most commonly in the form of sand.

SILICA GLASS: Glass made exclusively of silica.

SINTER: The process of heating that makes grains of a ceramic or metal a solid mass before it becomes molten.

SIZE: A glue, varnish, resin, or similar very dilute adhesive sealant used to block up the pores in porous surfaces or, for example, plaster and paper. Once the size has dried, paint or other surface coatings can be applied without the coating sinking in.

SLAG: A mixture of substances that are waste products of a furnace. Most slag are mainly composed of silicates.

SMELTING: Roasting a substance in order to extract the metal contained in it.

SODA: A flux for glassmaking consisting of sodium carbonate.

SOFTWOOD: Wood obtained from a coniferous tree.

SOLID: A rigid form of matter that maintains its shape regardless of whether or not it is in a container.

SOLIDIFICATION: Changing from a liquid to a solid.

SOLUBILITY: The maximum amount of a substance that can be contained in a solvent.

SOLUBLE: Readily dissolvable in a solvent.

SOLUTION: A mixture of a liquid (the solvent) and at least one other substance of lesser abundance (the solute). Like all mixtures, solutions can be separated by physical means.

SOLVAY PROCESS: Modern method of manufacturing the industrial alkali sodium carbonate (soda ash).

SOLVENT: The main substance in a solution.

SPECTRUM: A progressive series arranged in order, for example, the range of colors that make up visible light as seen in a rainbow.

SPINNERET: A small metal nozzle perforated with many small holes through which a filament solution is forced. The filaments that emerge are solidified by cooling and the filaments twisted together to form a yarn.

SPINNING: The process of drawing out and twisting short fibers, for example, wool, and thus making a thread or yarn.

SPRING: A natural flow of water from the ground.

STABILIZER: A chemical that, when added to other chemicals, prevents further reactions. For example, in soda lime glass the lime acts as a stabilizer for the silica.

STAPLE: A short fiber that has to be twisted with other fibers (spun) in order to make a long thread or yarn.

STARCHES: One form of carbohydrate. Starches can be used to make adhesives.

STATE OF MATTER: The physical form of matter. There are three states of matter: liquid, solid, and gas.

STEAM: Water vapor at the boiling point of water.

STONEWARE: Nonwhite pottery that has been fired at a high temperature until some of the clay has fused, a state called vitrified. Vitrification makes the pottery impervious to water. It is used for general tableware, often for breakfast crockery.

STRAND: When a number of yarns are twisted together, they make a strand. Strands twisted together make a rope.

SUBSTANCE: A type of material including mixtures.

SULFIDE: A compound that is composed only of metal and sulfur atoms, for example, PbS, the mineral galena.

SUPERCONDUCTORS: Materials that will conduct electricity with virtually no resistance if they are cooled to temperatures close to absolute zero (−273°C).

SURFACE TENSION: The force that operates on the surface of a liquid, and that makes it act as though it were covered with an invisible elastic film.

SURFACTANT: A substance that acts on a surface, such as a detergent.

SUSPENDED, SUSPENSION: Tiny particles in a liquid or a gas that do not settle out with time.

SYNTHETIC: Something that does not occur naturally but has to be manufactured. Synthetics are often produced from materials that do not occur in nature, for example, from petrochemicals. (i) Dye—a synthetic dye is made from petrochemicals, as opposed to natural dyes that are made of extracts of plants. (ii) Fiber—synthetic is a subdivision of artificial. Although both polyester and rayon are artificial fibers, rayon is made from reconstituted natural cellulose fibers and so is not synthetic, while polyester is made from petrochemicals and so is a synthetic fiber.

TANNIN: A group of pale-yellow or light-brown substances derived from plants that are used in dyeing fabric and making ink. Tannins are soluble in water and produce dark-blue or dark-green solutions when added to iron compounds.

TARNISH: A coating that develops as a result of the reaction between a metal and the substances in the air. The most common form of tarnishing is a very thin transparent oxide coating, such as occurs on aluminum. Sulfur compounds in the air make silver tarnish black.

TEMPER: To moderate or to make stronger: used in the metal industry to describe softening hardened steel or cast iron by reheating at a lower temperature or to describe hardening steel by reheating and cooling in oil; or in the glass industry, to describe toughening glass by first heating it and then slowly cooling it.

TEMPORARILY HARD WATER: Hard water that contains dissolved substances that can be removed by boiling.

TENSILE (PULLING STRENGTH): The greatest lengthwise (pulling) stress a substance can bear without tearing apart.

TENSION: A state of being pulled. Compare to compression.

TERRA COTTA: Red earth-colored glazed or unglazed fired clay whose origins lie in the Mediterranean region of Europe.

THERMOPLASTIC: A plastic that will soften and can be molded repeatedly into different shapes. It will then set into the molded shape as it cools.

THERMOSET: A plastic that will set into a molded shape as it first cools, but that cannot be made soft again by reheating.

THREAD: A long length of filament, group of filaments twisted together, or a long length of short fibers that have been spun and twisted into a continuous strand.

TIMBER: A general term for wood suitable for building or for carpentry and consisting of roughcut planks. *Compare to* **LUMBER**

TRANSITION METALS: Any of the group of metallic elements (for example, chromium and iron) that belong to the central part of the periodic table of the elements and whose oxides commonly occur in a variety of colors.

TRANSPARENT: Something that will readily let light through, for example, window glass. Compare to translucent, when only some light gets through but an image cannot be seen, for example, greaseproof paper.

TROPOSPHERE: The lower part of the atmosphere in which clouds form. In general, temperature decreases with height.

TRUNK: The main stem of a tree.

VACUUM: Something from which all air has been removed.

VAPOR: The gaseous phase of a substance that is a liquid or a solid at that temperature, for example, water vapor is the gaseous form of water.

VAPORIZE: To change from a liquid to a gas, or vapor.

VENEER: A thin sheet of highly decorative wood that is applied to cheap wood or engineered wood products to improve their appearance and value.

VINYL: Often used as a general name for plastic. Strictly, vinyls are polymers derived from ethylene by removal of one hydrogen atom, for example, PVC, polyvinylchloride.

VISCOSE: A yellow-brown solution made by treating cellulose with alkali solution and carbon disulfide and used to make rayon.

VISCOUS, VISCOSITY: Sticky. Viscosity is a measure of the resistance of a liquid to flow. The higher the viscosity—the more viscous it is—the less easily it will flow.

VITREOUS CHINA: A translucent form of china or porcelain.

VITRIFICATION: To heat until a substance changes into a glassy form and fuses together.

VOLATILE: Readily forms a gas. Some parts of a liquid mixture are often volatile, as is the case for crude oil. This allows them to be separated by distillation.

WATER CYCLE: The continual interchange of water between the oceans, the air, clouds, rain, rivers, ice sheets, soil, and rocks.

WATER VAPOR: The gaseous form of water.

WAVELENGTH: The distance between adjacent crests on a wave. Shorter wavelengths have smaller distances between crests than longer wavelengths.

WAX: Substances of animal, plant, mineral, or synthetic origin that are similar to fats but are less greasy and harder. They form hard films that can be polished.

WEAVING: A way of making a fabric by passing two sets of yarns through one another at right angles to make a kind of tight meshed net with no spaces between the yarns.

WELDING: Technique used for joining metal pieces through intense localized heat. Welding often involves the use of a joining metal such as a rod of steel used to attach steel pieces (arc welding).

WETTING: In adhesive spreading, a term that refers to the complete coverage of an adhesive over a surface.

WETTING AGENT: A substance that is able to cover a surface completely with a film of liquid. It is a substance with a very low surface tension.

WHITE GLASS: Also known as milk glass, it is an opaque white glass that was originally made in Venice and meant to look like porcelain.

WROUGHT IRON: A form of iron that is relatively soft and can be bent without breaking. It contains less than 0.1% carbon.

YARN: A strand of fibers twisted together and used to make textiles.

Set Index

USING THE SET INDEX

This index covers all nine volumes in the *Materials Science* set:

Volume
number   Title

1:    Plastics
2:    Metals
3:    Wood and paper
4:    Ceramics
5:    Glass
6:    Dyes, paints, and adhesives
7:    Fibers
8:    Water
9:    Air

An example entry:
Index entries are listed alphabetically.

sinter, sintering **2**: 21; **4**: 9, 44

Volume numbers are in bold and are followed by page references.

In the example above, "sinter, sintering" appears in Volume 2: Metals on page 21 and in Volume 4: Ceramics on pages 9 and 44. Many terms also are covered in the Glossary on pages 58–64.

See or *see also* refers to another entry where there will be additional relevant information.

A

abrasive **4**: 6, 12
ABS. *See* acrylonitrile-butadiene-styrene
acetate fiber **6**: 21; **7**: 36, 46
acetate film **1**: 47
acetic acid **1**: 37, 55
acid rain **8**: 57; **9**: 21, 46, 47
acidic water **8**: 6, 7, 46, 48, 52, 57
acids **1**: 15; **2**: 28, 30; **8**: 6, 46, 47, 48, 52, 56, 57
acrylic **1**: 38, 39, 40, 41
acrylic adhesives **6**: 50
acrylic fiber **1**: 39; **6**: 20, 21; **7**: 33, 36, 37, 38, 44, 45, 57
acrylic paints and stains **1**: 41; **6**: 32, 34, 35
acrylic plastics **1**: 38-41
acrylic powders **6**: 40
acrylonitrile-butadiene-styrene (ABS) **1**: 38
addition polymers/addition polymerization **1**: 10, 11, 27, 43; **7**: 15
additives **1**: 15, 16, 17; **3**: 51
adhesion **6**: 44, 45, 46
adhesives **1**: 22, 37, 40, 41, 42, 44, 53, 55; **3**: 8, 24, 43, 44, 45, 47, 50, 53, 54; **4**: 35, 41; **5**: 54; **6**: 4, 41-57

adhesive tapes **6**: 54, 57
admiralty brass **2**: 24
adobe **4**: 10, 11
advanced ceramics **4**: 42-57
aggregate **4**: 39, 41
air **9**: 4 AND THROUGHOUT
air bags **9**: 42
air brakes **9**: 35
air conditioning **9**: 26, 52
aircraft **2**: 21, 26, 27, 35, 51; **9**: 29, 32, 34, 35
air cushions **9**: 34-35
air drying **3**: 36
air gun **9**: 35
air in transportation **9**: 32
air pollution **9**: 19, 38-40, 44, 46-47
air pressure **9**: 5, 6, 28, 32, 37
albumen **6**: 49
alcohols **8**: 45, 51
alizarin **6**: 12, 13, 14
alkalis **1**: 15; **2**: 28, 30; **8**: 52
alkyd-based paint **6**: 31, 33
alkyd-based varnishes **6**: 37
alloys, alloying **1**: 15; **2**: 6, 13, 22, 23-27, 28, 34, 35, 37, 42; **4**: 46
alum **3**: 53; **6**: 10
alumina **4**: 38, 46, 50, 51, 54, 56, 57; **5**: 8, 9, 10, 13, 18, 52
aluminosilicates **4**: 14
aluminum **2**: 4, 5, 9, 10, 18, 19, 20, 21, 23, 24, 26, 27, 29, 30, 32, 50, 53; **4**: 14, 36
aluminum oxide **4**: 46, 50, 57; **5**: 13
amalgams **4**: 55
amides **7**: 10, 47
ammonia **9**: 41
amorphous solid **5**: 5, 15
amphibious vehicles **9**: 33
anaerobics **6**: 50
ancient glass **5**: 29
angle of incidence **5**: 20
aniline dyes **6**: 14, 22; **7**: 38
aniline mauve **6**: 14
animal glue **6**: 49
anions **8**: 10
annealing **5**: 50
anodized duralumin **2**: 32
anodizing **2**: 27, 32
antimony **2**: 45
antirust paint **6**: 33
anvil **2**: 12, 20
aqueous solutions **8**: 43, 44, 46
Araldite® **1**: 55
aramids **7**: 36, 50, 51
Archimedes' principle **8**: 38
argon **9**: 18, 36, 54, 55
armor **2**: 42, 43
armor plating **2**: 42
Arnel® **1**: 47; **7**: 46
arsenic oxide **5**: 11
artifacts **4**: 12
artificial dyes **6**: 7
artificial fibers **3**: 50; **7**: 7, 8, 9, 10, 12, 15, 16, 17, 19, 24, 30, 31, 32-57
artificial polymers **7**: 12
aspen **3**: 15
atmosphere **9**: 12, 14, 18, 20-21, 43, 44, 54, 55
atmospheric pressure **8**: 21, 22, 28; **9**: 6. *See also* air pressure
atomizer **9**: 28

atoms **2**: 6, 8, 9, 22, 23; **4**: 5, 9; **5**: 4, 5, 39; **7**: 4, 9; **8**: 8; **9**: 8, 10
atoms, properties in plastics **1**: 13
ax **3**: 6
azo dyes and pigments **6**: 7, 10; **7**: 38

B

backbone chain **1**: 27, 43, 55; **7**: 4. *See also* polymers
backbone unit **1**: 54. *See also* monomers
bagasse **3**: 49
Bakelite **1**: 43
balloons **9**: 8, 14, 51, 54
balsa **3**: 17, 20, 23
bamboo **3**: 49
band saw **3**: 34
barbed wire **2**: 6, 30, 31
barium carbonate **4**: 46; **5**: 9
barium titanate **4**: 46
bark **3**: 4, 6, 13, 14, 32
barometer **8**: 28
base metal **2**: 23, 24
bast fibers **7**: 20
batch kiln **4**: 19, 28
batch processing **7**: 32
batik **6**: 19
bauxite **4**: 38
beating metals **2**: 22-23
beech **3**: 17, 18, 23
bellows **9**: 28
bells **2**: 14, 44
bending ceramics **4**: 9
bending metals **2**: 12, 22, 35, 51
bends, the **9**: 42
benzene **1**: 33, 39
benzene ring **1**: 10; **6**: 15
Bessemer converter **2**: 46
Bessemer, Henry **2**: 47
binder **4**: 55; **6**: 27, 39
bioceramics **4**: 54-56
blacksmith **2**: 12, 22, 41
blast furnace **2**: 47
bleaches **6**: 24, 26
bleaching paper **3**: 52, 57
blending fibers **7**: 12, 25, 41, 43, 44, 45
blends, blending ceramics **4**: 17, 22, 36, 38
blood **8**: 6, 48
blood glue **6**: 49
bloom **2**: 40
blow molding **1**: 19
blown glass **5**: 32-33
board **3**: 34, 36, 42, 43, 44, 45, 46
bobbin **7**: 25, 42
boil, boiling water **8**: 11, 16, 20, 48
boilers **8**: 21, 22, 32, 33, 54
boiling point of water 11, 19, 20, 49, 48, 54
boll, boll fiber **7**: 4, 20, 25
bond paper **3**: 55
bonds and bonding **2**: 6, 7; **4**: 4, 5, 6, 9, 15, 25; **8**: 8, 9, 11, 14. *See also* covalent bonding, hydrogen bonds and bonding, ionic bonding
bone china **4**: 25
book paper **3**: 55
borax **5**: 13
boric oxide **5**: 8, 13
borosilicate glass **5**: 12, 13, 19
bottles **5**: 10, 28, 30, 43, 46-47
Boyle's law **9**: 8, 9
brass **2**: 6, 16, 24, 34, 41, 44